*SIEGEL — Topics in Complex Function Theory
Volume 1 — Elliptic Functions and Uniformization Theory
Volume 2 — Automorphic Functions and Abelian Integrals
Volume 3 — Abelian Functions and Modular Functions of Several Variables
STAKGOLD — Green's Functions and Boundary Value Problems
*STOKER — Differential Geometry
*STOKER — Nonlinear Vibrations in Mechanical and Electrical Systems
*STOKER — Water Waves: The Mathematical Theory with Applications
WESSELING — An Introduction to Multigrid Methods
WHITHAM — Linear and Nonlinear Waves
ZAUDERER — Partial Differential Equatiions of Applied Mathematics, 2nd Edition

*Now available in a lower priced paperback edition in the Wiley Classics Library

Finite Sums Decompositions in Mathematical Analysis

Finite Sums Decompositions in Mathematical Analysis

Themistocles M. Rassias
Athens, Greece
and
Jaromír Šimša
Brno, Czech Republic

JOHN WILEY & SONS
Chichester · New York · Brisbane · Toronto · Singapore

Baffins Lane, Chichester
West Sussex PO19 1UD, England
National 01243 779777
International (+44) 1243 779777

Other Wiley Editorial Offices

John Wiley & Sons, Inc., 605 Third Avenue,
New York, NY 10158-0012, USA

Jacaranda Wiley Ltd, 33 Park Road, Milton,
Queensland 4064, Australia

John Wiley & Sons (Canada) Ltd, 22 Worcester Road,
Rexdale, Ontario M9W 1L1, Canada

John Wiley & Sons (SEA) Pte Ltd, 37 Jalan Pemimpin #05-04,
Block B, Union Industrial Building, Singapore 2057

British Library Cataloguing in Publication Data

A catalogue record for this book is available from the British Library

ISBN 0 471 94827 6

Produced from camera-ready copy supplied by the authors using AMS-TeX
Printed and bound in Great Britain by Biddles Ltd, Guildford and King's Lynn

CONTENTS

PROLOGUE

Questions of representation of functions in several variables by means of functions of a smaller number of variables have captured the interest of mathematicians for centuries. One of these questions is closely connected with the thirteenth problem of D. Hilbert (1862–1943) and concerns the solvability of algebraic equations (see [H]). Let us mention the surprising result of A. N. Kolmogorov here: each continuous function h on the unit n-dimensional cube can be represented in the form

$$h(x_1, x_2, \ldots, x_n) = \sum_{i=1}^{2n+1} \varphi_i\Big(\sum_{j=1}^{n} \alpha_{ij}(x_j)\Big)$$

with some continuous functions φ_i and α_{ij}. Moreover, the inner functions α_{ij} can be chosen in advance, i.e. independently of the function h (see [Ko]).

Functions of certain special forms have been investigated by several authors, including J. d'Alembert (1717–1783), who as early as 1747 proved that each sufficiently smooth scalar function h of the form

$$h(x,y) = f(x) \cdot g(y) \tag{0.1}$$

has to satisfy the following partial differential equation

$$\frac{\partial^2 \log h}{\partial x \partial y} = 0$$

(see [d'A]). This equation can be also expressed in the form

$$\begin{vmatrix} h & h_y \\ h_x & h_{xy} \end{vmatrix} = 0\,. \tag{0.2}$$

A generalization of (0.1) to a finite sum of products of functions in single variables of the form

$$h(x,y) = \sum_{i=1}^{n} f_i(x) g_i(y) \tag{0.3}$$

has been considered since the beginning of this century. This forms the fundamental problem in the subject and the main undertaking of this book. The functions of the tensor product (0.3) play a significant role in many areas of both pure and applied mathematics; for example, in the theory of integral equations, ordinary and partial differential equations and approximation theory. In the year 1904 in the section *Arithmetics and Algebra* at the Third International Congress of Mathematicians in Heidelberg, Cyparissos Stéphanos announced the following result:
Functions of the type (0.3) *form the space of all solutions of the partial differential equation with the "Wronskian" of order* $(n+1)$

$$\det W_{n+1}h = \begin{vmatrix} h & h_y & \dots & h_{y^n} \\ h_x & h_{xy} & \dots & h_{xy^n} \\ \vdots & \vdots & \ddots & \vdots \\ h_{x^n} & h_{x^n y} & \dots & h_{x^n y^n} \end{vmatrix} = 0\,. \tag{0.4}$$

His talk was published in [St 1] (see also [St 2]), where some further applications and consequences were discussed. However, no proof of the above result was given and no smoothness condition on the given function h was mentioned. It appears that C. Stéphanos was thinking of analytic functions only. In fact, some 80 years later, Th. M. Rassias gave in [Ra 1] an example of a function in two real variables satisfying the differential equation (0.2), which is not of the form (0.1). For a short account of the development of the subject see [NR].

The new interest in the fundamental problem was revived early in the 1980's, when F. Neuman proved the basic theorem involving the equation (0.4) for functions of class C^n. This important result was published in [N 1] (see also a more available paper [N 2]), where also a general criterion for a function $h : X \times Y \to \mathbb{R}$ (or $\mathbb{C}$, respectively) being of the form (0.3) was given in the form of the functional equation

$$\begin{vmatrix} h(x_1,y_1) & h(x_1,y_2) & \dots & h(x_1,y_{n+1}) \\ h(x_2,y_1) & h(x_2,y_2) & \dots & h(x_2,y_{n+1}) \\ \vdots & \vdots & \ddots & \vdots \\ h(x_{n+1},y_1) & h(x_{n+1},y_2) & \dots & h(x_{n+1},y_{n+1}) \end{vmatrix} = 0$$

for all $x_i \in X$ and $y_j \in Y$. For the proof of this criterion no structure on the sets X and Y is required. (Most of the theorems here hold for both real- and complex-valued functions. This is why in this book we use $\mathbb{K}$ for the field $\mathbb{R}$ (reals) or for the field $\mathbb{C}$ (complex numbers).)

The problem of finding a necessary and a sufficient condition for a function h in several (more than two) variables to be represented by

$$h(x_1,x_2,\dots,x_k) = \sum_{i=1}^{n} f_{i1}(x_1)\cdot f_{i2}(x_2)\cdot \ldots \cdot f_{ik}(x_k) \tag{0.5}$$

was proposed by Th. M. Rassias in [Ra 2].

H. Gauchman and L. A. Rubel [GR] obtained some new results and extensions on finite sums expansions of the form (0.3), especially for real analytic functions.

The first existence theorem on the decomposition (0.5) was discovered by F. Neuman in [N 4]. Later, M. Čadek and J. Šimša in [ČŠ 1] found an effective criterion for the existence of the decomposition (0.5) by making use of a system of functional equations, which does not require any assumption on the function h. Furthermore, in [ČŠ 2] they outlined a way to find systems of partial differential equations whose solution space form the family of all sufficiently smooth functions h of type (0.5). Šimša [Ši 1] in turn has introduced some new functional equations for functions of the form (0.3) using the so-called Casorati determinant.

For a given function h that cannot be expressed in the form (0.3) with a prescribed value of n, it is an interesting problem to find the *best approximation* of the type

$$h(x,y) \approx \sum_{k=1}^{n} f_k(x) g_k(y) \tag{0.6}$$

in a given metric functional space. For the case of L^2 spaces, this problem was recently solved by Šimša [Ši 2]. However, for spaces without an inner product structure, the problem seems to be difficult and open. A weaker version of the above approximation problem, namely to find a good approximation (0.6) and also an estimation of the error, was treated in [Ši 3]. We mention here that in a related direction, the book of W. A. Light and E. W. Cheney [LC], published in 1985, is concerned with the approximation of multivariate functions by combinations of univariate ones.

Let us now review the contents of this book, chapter by chapter.

Chapter 1 introduces the functional determinants of Wronski, Gram and Casorati. Proofs of main properties of these determinants are given, and their relevance to the problem of the decomposition of (0.3) is described. Some classical results are combined with recent ones concerning the characterization of a function in several variables that can be represented in the form of the Casorati determinant of a system of functions in single variables.

In Chapter 2, we prove basic decomposition theorems. In fact, we solve the fundamental problem of finding necessary and sufficient conditions for a given function h in two variables (say x and y) to be decomposed in the form (0.3). We consider the main theorems about equation (0.4) due to F. Neuman, Th. M. Rassias, H. Gauchman and L. A. Rubel during the last decade. Neuman proved in [N 1]:

Let I and J be two intervals in $\mathbb{R}$ *and let* $h : I \times J \to \mathbb{K}$ *have the continuous derivative* $h_{x^n y^n}$ *on* $I \times J$. *If* h *is of the form* (0.3), *then*

$$\det W_{n+1} h(x,y) = 0 \quad \textit{for each point } (x,y) \in I \times J\,. \tag{0.7}$$

Conversely, if h satisfies (0.7) *and if in addition*

$$\det W_n h(x,y) \neq 0 \quad \textit{for each point } (x,y) \in I \times J, \tag{0.8}$$

then h is of the form (0.3) *on $I \times J$ with components $f_i \in C^n(I)$ and $g_i \in C^n(J)$.*

Rassias [Ra 1] obtained independently from F. Neuman a *local existence* version of the above result (see also [Ra 3]). Gauchman and Rubel [GR] considered the equation (0.7) for the class of real analytic functions h. Čadek and Šimša [ČŠ 2] found that under Neuman's condition above, the following product matrix formula holds

$$\begin{aligned} h(x,y) = {} & \big(h(x,y_0), h_y(x,y_0), \ldots, h_{y^{n-1}}(x,y_0)\big) \\ & \times W_n^{-1} h(x_0,y_0) \cdot \begin{pmatrix} h(x_0,y) \\ h_x(x_0,y) \\ \vdots \\ h_{x^{n-1}}(x_0,y) \end{pmatrix}. \end{aligned} \tag{0.9}$$

The next part of Chapter 2 is mainly devoted to decomposition criteria of F. Neuman and of J. Šimša involving the Casorati determinants

$$\det C_n h(\boldsymbol{x},\boldsymbol{y}) = \det \begin{pmatrix} h(x_1,y_1) & h(x_1,y_2) & \ldots & h(x_1,y_n) \\ h(x_2,y_1) & h(x_2,y_2) & \ldots & h(x_2,y_n) \\ \vdots & \vdots & \ddots & \vdots \\ h(x_n,y_1) & h(x_n,y_2) & \ldots & h(x_n,y_n) \end{pmatrix}$$

where $\boldsymbol{x} = (x_1, \ldots, x_n)$ and $\boldsymbol{y} = (y_1, \ldots, y_n)$.

Chapter 2 ends with a solution of the following matrix factorization problem

$$H(x,y) = F(x) \cdot G(y) \quad (x \in X,\ y \in Y) \tag{0.10}$$

in the group $GL_n(\mathbb{K})$ of all $n \times n$ nonsingular matrices with entries from the field $\mathbb{K}$. We show that all matrix functions of type (0.10) are exactly the solutions of the matrix functional equation

$$H(x,y) = H(x,y_1) \cdot H^{-1}(x_1,y_1) \cdot H(x_1,y)$$

and, in the case of functions H with continuous mixed partial derivative defined on a rectangle, these matrix functions are exactly the solutions of the matrix partial differential equation

$$H_{xy} = H_x \cdot H^{-1} \cdot H_y \,.$$

In Chapter 3 we solve the generalized problem of the decomposition of functions in several (more than two) variables that was posed by Rassias [Ra 2] and independently by Gauchman and Rubel [GR]. Namely, we establish different types of necessary and sufficient conditions (stated in terms of the Casorati and the partial Wronski determinants) for a function h to be expressed in the form

$$h(x,y,z) = \sum_{i=1}^{m}\sum_{j=1}^{n}\sum_{k=1}^{p} \alpha_{ijk} e_i(x) f_j(y) g_k(z) \tag{0.11}$$

or, for the case of functions in more than three variables,

$$h(x_1,\dots,x_k) = \sum_{i_1=1}^{m_1} \cdots \sum_{i_k=1}^{m_k} \alpha_{i_1\dots i_k} f^1_{i_1}(x_1)\cdot\ldots\cdot f^k_{i_k}(x_k)\,. \tag{0.12}$$

The idea of the proofs depends upon a reduction argument which brings the problem concerning functions for $k\ (\geq 3)$ variables to a system of k problems for functions in two vector variables.

We introduce the concept of the partial Wronski determinant and use it to obtain some information about the representations (0.11) and (0.12). We conclude Chapter 3 with the problem of matrix factorization

$$H(x_1,x_2,\dots,x_k) = H(x_1)\cdot H(x_2)\cdot\ldots\cdot H(x_k)\,. \tag{0.13}$$

In this direction, we show that the nonsingular matrix functions H of the form (0.13) are exactly all solutions of the following matrix functional equation

$$\begin{aligned} H(x_1,x_2,\dots,x_k) =& H(x_1,u_2,\dots,u_k)\cdot H_0^{-1}\cdot H(u_1,x_2,u_3,\dots,u_k)\cdot H_0^{-1}\cdot\ldots \\ &\ldots\cdot H_0^{-1}\cdot H(u_1,\dots,u_{k-1},x_k) \end{aligned}$$

where H_0^{-1} denotes the inverse of $H(u_1,\dots,u_k)$. A differential criterion for a smooth function H to be of type (0.13) can be stated in the form of the system of $\binom{k}{2}$ equations

$$H_{x_ix_j} = H_{x_i}\cdot H^{-1}\cdot H_{x_j} \quad (1\leq i<j\leq k)\,.$$

In Chapter 4 we develop a method for finding systems of linear partial differential equations whose solutions form a prescribed finite-dimensional space S of smooth functions in several variables. Namely,

$$S = \{f \mid f = \alpha_1 f_1 + \alpha_2 f_2 + \cdots + \alpha_n f_n : \alpha_1,\alpha_2,\dots,\alpha_n \in \mathbb{K}\}$$

where the basis $\{f_1,\dots,f_n\}$ forms an n-tuple of functions in $C^\infty(X,\mathbb{K})$ and X is a *region* in the k-dimensional space $\mathbb{R}^k$. Following [ČŠ 1], we show that if an

n-tuple of partial derivatives $D_1 = \mathrm{id}, D_2, \ldots, D_n$ is chosen in such a way that the following general Wronski determinant

$$\det W_n[D_1, \ldots, D_n; f_1, \ldots, f_n] = \det \begin{pmatrix} D_1 f_1 & D_1 f_2 & \ldots & D_1 f_n \\ D_2 f_1 & D_2 f_2 & \ldots & D_2 f_n \\ \vdots & \vdots & \ddots & \vdots \\ D_n f_1 & D_n f_2 & \ldots & D_n f_n \end{pmatrix} \tag{0.14}$$

is nonvanishing on the whole region X, then the space S consists exactly of all solutions of the system

$$\left(\frac{\partial}{\partial x_i} \circ D_r\right) f = \sum_{s=1}^{n} a_{irs}(x) \cdot D_s f \quad (i = 1, 2, \ldots, k;\ r = 1, 2, \ldots, n) \tag{0.15}$$

where $x_1, \ldots, x_n$ are coordinates in $\mathbb{R}^k$. We analyse an algorithm which enables one to find the desired system of partial derivatives needed for the nonvanishing determinant (0.14). We also discuss the problem of reducing the system (0.15) to a system with a lesser number of equations.

In Chapter 5, we present necessary and sufficient conditions for a smooth function $h \in C^\infty(X \times Y, \mathbb{K})$ in two *vector* variables x and y to be represented in the form (0.3). Here X and Y are two regions in $\mathbb{R}^p$ and $\mathbb{R}^q$, respectively. This is done by means of the general Wronski matrix of the function h defined by

$$W_n[D_i; d_j]h = \begin{pmatrix} D_1 \circ d_1 h & D_1 \circ d_2 h & \ldots & D_1 \circ d_n h \\ D_2 \circ d_1 h & D_2 \circ d_2 h & \ldots & D_2 \circ d_n h \\ \vdots & \vdots & \ddots & \vdots \\ D_n \circ d_1 h & D_n \circ d_2 h & \ldots & D_n \circ d_n h \end{pmatrix} \quad (n = 1, 2, \ldots)$$

where $D_1, \ldots, D_n$ and $d_1, \ldots, d_n$ are partial derivatives in the corresponding coordinate spaces. We also study the global decomposition problem (0.3) for functions h defined on the Cartesian product of two topological manifolds. The following result indicates a condition under which local decompositions (0.3) of a function h can be glued to a global one.

Let X and Y be two arcwise connected topological spaces. Suppose that a function $h : X \times Y \to \mathbb{K}$ has the following property: Each point $(x, y) \in X \times Y$ has a neighbourhood $U \times V \subseteq X \times Y$ in which the function h is of the form (0.3), where $n \geq 1$ is a given integer and $f_i : U \to \mathbb{K}$ and $g_i : V \to \mathbb{K}$ are locally linearly independent functions. Then h is of the form (0.3) *on the whole set $X \times Y$.*

In Chapter 6, we introduce approximate decompositions for the case of smooth functions in two real variables which cannot be represented in the form (0.3) and we also estimate the errors in the corresponding formulas. It is shown that Neuman's conditions, stated in Chapter 2, are *stable* in the following sense. If I and J are compact intervals in $\mathbb{R}$ and if a function h satisfies

$$\det W_n h(x,y) \neq 0 \quad \text{and} \quad \left| \frac{\det W_{n+1} h(x,y)}{\det W_n h(x,y)} \right| \leq \varepsilon \quad (x \in I,\ y \in J)$$

with some constant $\varepsilon > 0$ and some integer $n \geq 1$, then

$$\big|h(x,y) - T(x,y)\big| \leq \frac{\varepsilon}{(n!)^2} \big(1 + K_1|x - x_0|\big)\big(1 + K_2|y - y_0|\big)\big|(x - x_0)(y - y_0)\big|^n \tag{0.16}$$

where K_1 and K_2 are suitable constants and the approximation function T is equal to the right-hand side of formula (0.9). We also prove the sharpness of formula (0.16) in the sense that the number ε in the right-hand side of (0.16) cannot be replaced by any smaller value $m(\varepsilon)$.

In Chapter 7, we deal with the best L^2-approximations of the form (0.6). The proof of the corresponding existence theorem is essentially given by the fact that the family of all right-hand sides of (0.6) with the prescribed n is closed in the weak topology of the Hilbert space $L^2(X \times Y)$, where both X and Y are both σ-additive and σ-finite measure spaces. To show this, we use the technique of the Casorati determinants for a bilinear function $B : L^2(X) \times L^2(Y) \to \mathbb{K}$ defined by

$$B(u,v) = \int_{X \times Y} h u \bar{v}\, d(\mu \times \lambda) \qquad (u \in L^2(X) \text{ and } v \in L^2(Y)).$$

Then we verify that components f_i and g_i of the best approximation (0.6) have to satisfy a certain system of $2n$ integral equations. Using basic properties from the spectral theory of self-adjoint linear operators on Hilbert spaces, we obtain a full description of the best approximations. As a corollary, we get the so-called Hilbert–Schmidt decomposition theorem. .

In last Chapter 8, we utilize a geometric framework for partial differential equations (see [Gos, KLV, Ly, LP, Pr 1, Pr 2, Pr 3, PR 1, PR 2, Vi]) in order to prove that the set of solutions of the d'Alembert equation (0.2) is larger than the set of smooth functions of the form (0.1). This agrees with the previous example of Th. M. Rassias mentioned above. The result of such a geometric approach is the following:

The set of two-dimensional integral manifolds of the equation (0.2) *properly contains the one representable by graphs of* 2*-jet-derivatives of functions* $h(x,y)$ *expressed in the functional form* (0.1).

A generalization of this result to functions of more than two variables is sketched also by considering the following generalized d'Alembert equation

$$\frac{\partial^k \log h}{\partial x_1 \partial x_2 \ldots \partial x_k} = 0$$

in which $h = h(x_1, x_2, \ldots, x_k)$ is a scalar unknown function, smoothly depending on the variables $x_1, \ldots, x_k$.

The aim of the present work is to give a self-contained presentation of the above-mentioned results with emphasis on the progress that has been achieved during the last decade, including some very recent unpublished results.

The methods that we employ combine ideas in Functional and Differential Equations as well as Differential Geometry.

This work provides the present stage of the theory that started at least 250 years ago and has undergone very exciting development in this century. We believe that several new results will be added to it in the future, as there are still many open problems (cf. the last section of the book) which continue to resist solution in this classical field of Mathematical Analysis.

We wish to express our thanks to Professor Gustave Choquet, Professor Valentin Lychagin, Professor František Neuman, Professor Agostino Prástaro, Dr Martin Čadek, Dr Jiří Vanžura and Dr Kenneth Wolsson, who read parts of the manuscript and provided some very useful comments. We point out here that in the writing of the book, the habilitation thesis [Ši 5] of the second author played a significant role.

We are also very much indebted to Dr Jan Slovák and Dr Karel Horák, who helped us to arrange the final copy of our manuscript in the TEX typesetting system. Finally, it is also our pleasure to acknowledge the fine cooperation and assistance that Mrs Helen Ramsey and Mr David Ireland of John Wiley & Sons have provided in the publication of this book.

July 1994

Themistocles M. Rassias
Jaromír Šimša

1 FUNCTIONAL DETERMINANTS

This chapter deals mainly with three types of functional determinants which enable one to state criteria of linear dependence in various functional spaces.

1.1. Linear dependence in functional spaces

A general approach for finding decompositions of the form

$$h(x,y) = \sum_{i=1}^{n} f_i(x) g_i(y) \tag{1.1.1}$$

is based on the following simple idea.

If L is an arbitrary linear operator or linear functional acting on the space of functions in the variable y, then the following function of the variable x

$$L(h(x,y)) = \sum_{i=1}^{n} f_i(x) L(g_i(y))$$

is an element of the linear finite-dimensional functional space generated by the n-tuple $f_1, \ldots, f_n$. Choosing $n+1$ such operators $L = L_1, \ldots, L_{n+1}$, we therefore obtain the system of functions

$$L_1(h(x,y)), L_2(h(x,y)), \ldots, L_{n+1}(h(x,y)) \tag{1.1.2}$$

which is linearly dependent (in the variable x). Moreover, if the operators L_i are chosen "well" in the sense that there exists a linearly independent n-tuple among the functions (1.1.2), *then each "component" f_i in the unknown decomposition* (1.1.1) *is a linear combination of functions from this n-tuple.*

Recall that we usually consider a family S of functions $f : X \to \mathbb{K}$ as a linear space over the field $\mathbb{K}$ with respect to the pointwise defined operations

$$(f+g)(x) := f(x) + g(x) \quad \text{and} \quad (\alpha \cdot f)(x) := \alpha \cdot f(x)$$

for all $f, g \in S$, $x \in X$ and $\alpha \in \mathbb{K}$. (Throughout the book, the letter $\mathbb{K}$ will denote any of the fields $\mathbb{R}$ or $\mathbb{C}$.) Hence, we say that a system of functions $f_i : X_i \to \mathbb{K}$ $(1 \le i \le n)$ is *linearly dependent* in the set X, if $X \subseteq X_i$ $(1 \le i \le n)$ and if there exist constants $\alpha_1, \alpha_2, \ldots, \alpha_n \in \mathbb{K}$ (not all equal to zero) satisfying

$$\alpha_1 f_1(x) + \alpha_2 f_2(x) + \cdots + \alpha_n f_n(x) = 0 \qquad \text{for each } x \in X\,. \tag{1.1.3}$$

The function in the left-hand side is called a *linear combination* of the given functions $f_1, \ldots, f_n$. Conversely, if (1.1.3) is satisfied with $\alpha_1 = \ldots = \alpha_n = 0$ only, then we say that the functions $f_1, f_2, \ldots, f_n$ are *linearly independent* in the set X. In the case when X is a topological space and the functions $f_1, \ldots, f_n$ are linearly independent in each nonempty open subset of X, we say that the functions $f_1, \ldots, f_n$ are *locally linearly independent*.

Let us mention an easy consequence of the *uniqueness theorem* in the theory of analytic functions: If an n-tuple of analytic functions is linearly independent in a region of the complex plane, then this n-tuple is locally linearly independent. The same conclusion holds also for n-tuples of real analytic functions defined in an interval on the real line. However, this conclusion fails to hold in general for functions of the classes C^∞, C^n or L^2.

1.2. The Wronski determinant

Let I be an interval in $\mathbb{R}$ and let $f_1, f_2, \ldots, f_n \in C^{n-1}(I, \mathbb{K})$. The following matrix of size $n \times n$

$$W(f_1, f_2, \ldots, f_n; t) := \begin{pmatrix} f_1(t) & f_2(t) & \ldots & f_n(t) \\ f_1'(t) & f_2'(t) & \ldots & f_n'(t) \\ \vdots & \vdots & \ddots & \vdots \\ f_1^{(n-1)}(t) & f_2^{(n-1)}(t) & \ldots & f_n^{(n-1)}(t) \end{pmatrix} \tag{1.2.1}$$

is called the *Wronski matrix* of the n-tuple of functions $f_1, \ldots, f_n$ (cf. [PS]). The value of its determinant $w(t) = \det W(f_1, f_2, \ldots, f_n; t)$ is called the *Wronski determinant* (or the *Wronskian*[1]) of the n-tuple $f_1, \ldots, f_n$. In this case, the Wronskian is a continuous mapping $w : I \to \mathbb{K}$. The basic property of the Wronski determinant is described in the following theorem.

[1]After the Polish mathematician known as Wronski, whose real name appears to have been Józef Maria Hoehne–Wrónski (1778–1853). He was the son of a Polish architect.

Theorem 1.2.1: *Let the functions $f_1, \dots, f_n \in C^{n-1}(I, \mathbb{K})$ be linearly dependent in the interval $I \subseteq \mathbb{R}$. Then the Wronskian of $f_1, \dots, f_n$ is equal to zero at each point of I. If in addition the Wronskian of $f_1, f_2, \dots, f_{n-1}$ is nonzero at each point of I, then the function f_n is a linear combination of $f_1, \dots, f_{n-1}$ in the set I.*

Proof: Since the functions $f_1, \dots, f_n \in C^{n-1}(I, \mathbb{K})$ are assumed to be linearly dependent in the interval I, there exist constants $\alpha_1, \dots, \alpha_n$, not all zero, such that

$$\alpha_1 f_1(t) + \alpha_2 f_2(t) + \cdots + \alpha_n f_n(t) = 0 \quad \text{for each } t \in I\,. \tag{1.2.2}$$

Differentiating this equality j times, we get

$$\alpha_1 f_1^{(j)}(t) + \alpha_2 f_2^{(j)}(t) + \cdots + \alpha_n f_n^{(j)}(t) = 0 \quad (t \in I) \tag{1.2.3}$$

for each $j = 1, 2, \dots, n-1$. We may consider (1.2.2) and the $(n-1)$ equations (1.2.3) as a linear algebraic system in $\alpha_1, \dots, \alpha_n$, with a parameter $t \in I$. Since this system has a nontrivial solution, its determinant (equal to the Wronskian of $f_1, \dots, f_n$) vanishes at each point $t \in I$. If in addition

$$\det W(f_1, f_2, \dots, f_{n-1}; t) = \begin{vmatrix} f_1(t) & f_2(t) & \dots & f_{n-1}(t) \\ f_1'(t) & f_2'(t) & \dots & f_{n-1}'(t) \\ \vdots & \vdots & \ddots & \vdots \\ f_1^{(n-2)}(t) & f_2^{(n-2)}(t) & \dots & f_{n-1}^{(n-2)}(t) \end{vmatrix} \neq 0 \quad \text{for each } t \in I\,, \tag{1.2.4}$$

then the first $n-1$ columns of the Wronski matrix $W(f_1, f_2, \dots, f_n; t)$ are linearly independent, for each (fixed) $t \in I$. Since the matrix $W(f_1, f_2, \dots, f_n; t)$ is assumed to be singular, we conclude that its last column is a linear combination of the preceding ones. So there exist $n-1$ coefficients $\alpha_i : I \to \mathbb{K}$ such that

$$f_n^{(j)}(t) = \sum_{i=1}^{n-1} \alpha_i(t) f_i^{(j)}(t) \quad (t \in I,\ 0 \le j \le n-1). \tag{1.2.5}$$

In view of (1.2.4), the values of α_i are uniquely determined from the first $n-1$ equations (1.2.5) with $j = 0, 1, \dots, n-2$. Moreover, Cramer's rule leads to the conclusion that all the functions $\alpha_i : I \to \mathbb{K}$ are differentiable at each point of I. Thus, we can write

$$\begin{aligned} \sum_{i=1}^{n-1} \alpha_i'(t) f_i^{(j)}(t) &= \frac{d}{dt}\left(\sum_{i=1}^{n-1} \alpha_i(t) f_i^{(j)}(t) \right) - \sum_{i=1}^{n-1} \alpha_i(t) f_i^{(j+1)}(t) \\ &= \frac{d}{dt}(f_n^j(t)) - f_n^{j+1}(t) = 0 \quad (t \in I) \end{aligned}$$

for each $j=0,1,\dots,n-2$. However, condition (1.2.4) implies that the system of $n-1$ equalities

$$\sum_{i=1}^{n-1} \alpha_i'(t) f_i^{(j)}(t) = 0 \quad (j=0,1,\dots,n-2)$$

holds only if $\alpha_i'(t)=0$ $(i=1,2,\dots,n-1,\ t\in I)$. Consequently, each function α_i is constant on I and then the relation (1.2.5) with $j=0$ implies that f_n is a linear combination of $f_1,\dots,f_{n-1}$ in the set I. □

Remarks 1.2.2: (i) Simple examples show that the Wronski determinant of an n-tuple $f_1,\dots,f_n \in C^{n-1}(I,\mathbb{K})$ can vanish at each point of the interval I, although the functions $f_1,\dots,f_n$ are linearly independent (in the whole interval I). However, if the functions $f_1,\dots,f_n \in C^{n-1}(I,\mathbb{K})$ are locally linearly independent, then their Wronskian does not vanish almost everywhere in I. To prove this, suppose on the contrary that the Wronskian of $f_1,\dots,f_n$ is equal to zero at each point of some subinterval $J \subseteq I$. Since $f_1,\dots,f_n$ are assumed to be linearly independent in J, we have $f_1(t_0) \neq 0$ for some $t_0 \in J$. Hence, there exists the smallest integer r $(1 < r \leq n)$ so that the Wronskian of the r-tuple $f_1,\dots,f_r$ vanishes at each point of J. Then the Wronskian of $f_1,\dots,f_{r-1}$ does not vanish at any point of some open subinterval $K \subseteq J$ and Theorem 1.2.1 therefore implies that f_r is a linear combination of $f_1,\dots,f_{r-1}$ in the set K, which contradicts the local linear independence of $f_1,\dots,f_n$.

(ii) The pathological case described in the first sentence of (i) is impossible, provided that the functions $f_1,\dots,f_n \in C^n(I,\mathbb{K})$ are solutions of the same linear differential equation of order n

$$f^{(n)} + p_1(t) f^{(n-1)} + \cdots + p_n(t) f = 0 \qquad (t \in I) \tag{1.2.6}$$

with continuous coefficients $p_i : I \to \mathbb{K}$. Indeed, it follows from the Liouville formula (cf. [He])

$$w(t) \cdot \exp \int_{t_0}^{t} p_1(s)\, ds = w(t_0) \qquad (t, t_0 \in I) \tag{1.2.7}$$

that the Wronskian w of an arbitrary n-tuple of solutions of (1.2.6) satisfies either $w(t) \neq 0$ $(t \in I)$, or $w(t)=0$ $(t \in I)$. A similar formula is due to Niels Henrik Abel (1802 – 1829). It was published in one of his five memoirs in Volume 2 (1827) of *Crelle's Journal*, known today as *Journal der Reine und Angewandte Mathematik*. The proof of (1.2.7) is quite elementary: we need only to show that the Wronskian w is a solution of the first-order differential equation

$$\frac{dw}{dt} = -p_1(t) w\,. \tag{1.2.8}$$

To verify this, let us differentiate the determinant w of the Wronski matrix (1.2.1) according the known rule "row by row". Using the fact that any determinant with two identical rows vanishes, we obtain the equality

$$w'(t) = \begin{vmatrix} f_1(t) & f_2(t) & \dots & f_n(t) \\ f_1'(t) & f_2'(t) & \dots & f_n'(t) \\ \vdots & \vdots & \ddots & \vdots \\ f_1^{(n-2)}(t) & f_2^{(n-2)}(t) & \dots & f_n^{(n-2)}(t) \\ f_1^{(n)}(t) & f_2^{(n)}(t) & \dots & f_n^{(n)}(t) \end{vmatrix}.$$

Substituting $f_i^{(n)}(t) = -\sum_{k=0}^{n-1} p_{n-k}(t) f_i^{(k)}(t)$ into the last row of this determinant, we get

$$w'(t) = -\sum_{k=0}^{n-1} p_{n-k}(t) \cdot \begin{vmatrix} f_1(t) & f_2(t) & \dots & f_n(t) \\ f_1'(t) & f_2'(t) & \dots & f_n'(t) \\ \vdots & \vdots & \ddots & \vdots \\ f_1^{(n-2)}(t) & f_2^{(n-2)}(t) & \dots & f_n^{(n-2)}(t) \\ f_1^{(k)}(t) & f_2^{(k)}(t) & \dots & f_n^{(k)}(t) \end{vmatrix}.$$

Since the determinant in the last sum vanishes for each $k < n-1$, we therefore conclude that the equation (1.2.8) holds and the proof of (1.2.7) is complete.

Let us add the converse fact that if the Wronskian w of some n-tuple of functions $f_1, \dots, f_n \in C^n(I, \mathbb{K})$ is nonzero at each point $t \in I$, then the functions $f_1, \dots, f_n$ form a fundamental set of solutions of the unique equation (1.2.6) which can be derived from

$$\frac{1}{w(t)} \cdot \det \begin{pmatrix} f_1(t) & f_2(t) & \dots & f_n(t) & f \\ f_1'(t) & f_2'(t) & \dots & f_n'(t) & f' \\ \vdots & \vdots & \ddots & \vdots & \vdots \\ f_1^{(n)}(t) & f_2^{(n)}(t) & \dots & f_n^{(n)}(t) & f^{(n)} \end{pmatrix} = 0$$

by expanding the determinant with respect to the last column. The uniqueness of (1.2.6) can be easily proved by using Cramer's rule for computing "unknown" coefficients $p_1, \dots, p_n$ from the system of n equations (1.2.6) with $f = f_1, \dots, f_n$.

(iii) If the functions $f_1, \dots, f_n$ are *real* or *complex analytic* on the interval I and if w denotes their Wronskian, then the condition

$$w(t) = 0 \qquad \text{for each } t \in I \tag{1.2.9}$$

is not only necessary (see Theorem 1.2.1) but also sufficient for $f_1, \dots, f_n$ to be linearly dependent in the whole interval I. To prove this result (cf. [C]), assume that

(1.2.9) holds and denote by r the largest positive integer such that the Wronskian of the r-tuple $f_1, \dots, f_r$ has no zero value in some nonempty open subinterval $J \subseteq I$. (The number r exists excepting the case when $f_1 = 0$ on I, which is an obvious case of the linear dependence of $f_1, \dots, f_n$.) In view of (1.2.9), we have $r \leq n-1$. Since the Wronskian of the $(r+1)$-tuple $f_1, \dots, f_{r+1}$ vanishes at each point of J, we may apply Theorem 1.2.1 to conclude that

$$f_{r+1}(t) = \gamma_1 f_1(t) + \gamma_2 f_2(t) + \cdots + \gamma_r f_r(t) \quad (t \in J)$$

with some constants γ_i. By virtue of the uniqueness theorem for analytic functions, the last equality must be valid at each point of the whole interval I, which proves the linear dependence of $f_1, \dots, f_n$ in I.

(iv) In Chapter 4 we will show the way in which the Wronski matrix (1.2.1) can be defined in the case when $f_1, \dots, f_n$ are scalar functions in one vector independent variable and how, in turn, to extend the applicability of Theorem 1.2.1 to this case (Theorem 4.1.1). For earlier mention of this type of extension, cf. [O], [Kr] and [W 2].

Let us continue our discussion about the linear dependence of an n-tuple of functions $f_1, \dots, f_n$ with the vanishing Wronskian. It is clear that Theorem 1.2.1 ensures the linear dependence of $f_1, \dots, f_n$ in the interval I just in the case when at least one of the n Wronskians

$$w_i(t) = \det W(f_1, \dots, f_{i-1}, f_{i+1}, \dots, f_n; t) \quad (i=1, \dots, n) \tag{1.2.10}$$

has no zero value in the whole interval I. A weaker form of this condition, namely

$$\sum_{i=1}^{n} |w_i(t)| \neq 0 \quad \text{for each } t \in I, \tag{1.2.11}$$

was proposed by G. Peano in [Pe]. He proved the following:

Theorem 1.2.3 [Pe]: *Let the Wronskian of $f_1, \dots, f_n$ vanish at each point of the interval I, and let the Wronskians* (1.2.10) *satisfy condition* (1.2.11). *Then the functions $f_1, \dots, f_n$ are linearly dependent in the whole interval I.*

Proof: Theorem 1.2.3 is an immediate consequence of Theorem 1.2.1, because of the following result of M. Bôcher.

Lemma 1.2.4 [Bôc]: *Under the hypotheses of Theorem 1.2.3, each Wronskian w_i satisfies either $w_i(t) \neq 0$ $(t \in I)$, or $w_i(t) = 0$ $(t \in I)$.*

To avoid unnecessary repetition, we omit the proof of Lemma 1.2.4 now, because a more general result, stated as Theorem 1.2.5, will be proved shortly.

In what follows, we describe the theory of critical points of function sets, developed by Wolsson [W 1]. He remarkably extended Bôcher's idea of Lemma 1.2.4 to obtain a final answer to the problem of a criterion of linear dependence stated in terms of the Wronski determinants.

Let us start with a short series of notations and definitions. Up to the end of this section, Φ will denote a fixed n-tuple of functions $f_1, \dots, f_n \in C^{n-1}(I, \mathbb{K})$, where I is a nonempty open interval of reals. The Wronskian of Φ will be denoted by $w[\Phi]$. Each point $t \in I$, at which $w[\Phi] = 0$, will be called the *critical point* (of Φ). Each Wronskian $w[\Psi]$, where Ψ is a r-element subset of Φ, will be called the *sub-Wronskian* (of Φ) of order r. The *order* of a critical point t of Φ is the largest positive integer r such that some sub-Wronskian (of Φ) of order r does not vanish at the point t. (Should all sub-Wronskians vanish at t, the order of t is defined to be zero.) We denote the order of a critical point t by $\operatorname{ord} t$.

Theorem 1.2.5 [W 1]: *Let I_r be an open subinterval of I consisting only of the critical points of Φ of the same order $r > 0$. Then the n-tuple Φ is linearly dependent in I_r. Moreover, the linear space spanned by Φ on I_r is r-dimensional and each sub-Wronskian $w[\Psi]$ of order r satisfies either $w[\Psi](t) = 0$ $(t \in I_r)$, or $w[\Psi](t) \neq 0$ $(t \in I_r)$, according to whether the r-tuple Ψ is linearly dependent or not.*

Before we give a proof of the above result, notice that Lemma 1.2.4 is identical with Theorem 1.2.5 for $r = n - 1$.

Proof: Take any point $t_0 \in I_r$. Since $\operatorname{ord} t_0 = r$, there exists an r-element subset $\Psi \subsetneq \Phi$ such that $w[\Psi]$ is nonzero at t_0 and so by continuity, in an open interval about t_0. Denote by $I' \subseteq I_r$ the largest open interval containing t_0 in which $w[\Psi] \neq 0$. We will show that $I' = I_r$. If $I' = (a, b)$ were different from I_r, one of its end point, say b, would belong to I_r. By maximality, $w[\Psi](b) = 0$. It is no loss of generality to assume that $\Psi = (f_1, \dots, f_r)$. For each $k = r + 1, \dots, n$, we now apply Theorem 1.2.1 to the $(r+1)$-tuple $\Psi' = \Psi \cup \{f_k\}$: For each $t \in I'$, we have $w(\Psi')(t) = 0$ and $w(\Psi)(t) \neq 0$, because $\operatorname{ord} t = r$. Therefore, the function f_k is a linear combination of Ψ in I':

$$f_k(t) = \sum_{i=1}^{r} \alpha_{ki} f_i(t) \quad (t \in I').$$

Differentiating $r - 1$ times yields the equalities

$$f_k^{(j)}(t) = \sum_{i=1}^{r} \alpha_{ki} f_i^{(j)}(t) \quad (t \in I')$$

for each $j = 0, 1, \dots, r-1$. By continuity, the last r equalities hold also at $t = b$.

Thus, each column of the $r \times n$ matrix

$$\begin{pmatrix} f_1(b) & f_2(b) & \dots & f_n(b) \\ f_1'(b) & f_2'(b) & \dots & f_n'(b) \\ \vdots & \vdots & \ddots & \vdots \\ f_1^{(r-1)}(b) & f_2^{(r-1)}(b) & \dots & f_n^{(r-1)}(b) \end{pmatrix} \tag{1.2.12}$$

is a linear combination of the first r ones. However, these r columns are linearly dependent, because of $w[\Psi](b) = 0$. Consequently, the rank of the matrix (1.1.12) does not exceed $r-1$, which contradicts to $\operatorname{ord} b = r$. This contradiction results from assuming the end point b lies inside I_r. (The case when the other end point a lies inside I_r can be treated analogously.) Thus, we have proved that $w[\Psi](t) \neq 0$ at each point $t \in I_r$, for each r-element subset $\Psi \subsetneq \Phi$ satisfying $w[\Psi](t_0) \neq 0$ at *some* point $t_0 \in I_r$. We have also established that such a Ψ forms a basis of the (r-dimensional) linear space spanned by Φ on I_r. The proof is complete. □

Suppose that the Wronskian $w[\Phi]$ vanishes at each point of the interval I and define its subsets $\mathscr{C}_r$ by

$$\mathscr{C}_r := \{t \in I \mid \operatorname{ord} t = r\} \quad \text{for } r = 0, 1, \dots, n-1\,.$$

Then

$$I = \mathscr{C}_0 \,\dot\cup\, \mathscr{C}_1 \cup \dots \cup \mathscr{C}_{n-1}$$

is written as a disjoint union of the subsets $\mathscr{C}_r$. Denote by $\operatorname{Int} \mathscr{C}_r$ the set of all inner points of $\mathscr{C}_r$ (the *interior* of $\mathscr{C}_r$). It is well known that $\operatorname{Int} \mathscr{C}_r$ (as any open subset of the real line) is either empty or can be represented as a union of a countable collection of disjoint open intervals, the so-called *component intervals*. Following K. Wolsson, we call each component interval of $\operatorname{Int} \mathscr{C}_r$ to be the *interval of dependence* of the n-tuple Φ. Theorem 1.2.5 ensures that the intervals of dependence are aptly named, and the following result (Theorem 1.2.6) ensures that they are dense everywhere in I. (Notice that Theorem 1.2.5 can be applied to each interval of dependence excepting the component intervals of $\operatorname{Int} \mathscr{C}_0$. However, if an interval I_0 consists of only critical points of order zero, then $f_1 = \ldots = f_n = 0$ at each point of I_0; hence the subinterval I_0 has no "influence" on the linear dependence of Φ in the whole interval I.)

Theorem 1.2.6 [W 1]: *Let the interval I consist only of the critical points of the n-tuple Φ. Then the collection of all intervals of dependence of Φ is dense in I, i.e. each point of I lies in the closure of the union of these intervals.*

Proof: To prove that $\bigcup_{r=0}^{n-1} \operatorname{Int} \mathscr{C}_r$ is dense in I, suppose on the contrary that

$$(a, b) \cap \operatorname{Int} \mathscr{C}_r \text{ is empty} \quad (r = 0, 1, \dots, n-1) \tag{1.2.13}$$

for some small but nonempty open subinterval $(a,b) \subset I$. Define

$$r_0 = \max\{\operatorname{ord} t \mid t \in (a,b)\}$$

and choose $t_0 \in (a,b)$ so that $\operatorname{ord} t_0 = r_0$. Then there exists a sub-Wronskian $w[\Psi]$ of order r_0 satisfying $w[\Psi](t_0) \neq 0$. By continuity, we have $w[\Psi](t) \neq 0$ at each point t of some open subinterval $I_0 \subseteq (a,b)$. Thus, $\operatorname{ord} t = r_0$ for each $t \in I_0$, which means that $I_0 \subseteq \operatorname{Int} \mathscr{C}_{r_0}$. This contradicts (1.2.13) and the proof is complete. □

Let $\{I_\alpha \mid \alpha \in A\}$ denote the system of all intervals of dependence of a given n-tuple Φ, where A is a countable index set. By definition, each interval I_α consists only of the critical points of the same order, say r_α. In view of Theorem 1.2.5, the collection

$$N_\alpha := \{\vec{c} = (\gamma_1, \dots, \gamma_n) \in \mathbb{K}^n \mid \gamma_1 f_1(t) + \cdots + \gamma_n f_n(t) = 0 \quad (t \in I_\alpha)\} \tag{1.2.14}$$

forms an $(n-r)$-dimensional linear subspace of $\mathbb{K}^n$. The final criterion of linear dependence is given by the following theorem. It shows that the linear dependence of Φ on I is equivalent to the condition that the intersection of all subspaces N_α is nontrivial.

Theorem 1.2.7 [W 1]: *Let Φ be an n-tuple of functions in $C^{n-1}(I, \mathbb{K})$ and let $\{I_\alpha \mid \alpha \in A\}$ denote the system of all intervals of dependence of Φ. If N_α are defined as in* (1.2.14)*, then the n-tuple Φ is linearly dependent in the interval I if and only if*

(i) $w[\Phi](t) = 0$ *(at each point $t \in I$)*

and

(ii) *the intersection $\bigcap_{\alpha \in A} N_\alpha$ is a nonzero subspace of $\mathbb{K}^n$.*

Proof: First note that the necessity of (i) follows from the basic result of this section, Theorem 1.2.1. If Φ is linearly dependent in I, then by definition,

$$\gamma_1 f_1(t) + \cdots + \gamma_n f_n(t) = 0 \quad (t \in I)$$

for some nonzero $\vec{c} = (\gamma_1, \dots, \gamma_n) \in \mathbb{K}^n$. Such a vector $\vec{c}$ lies in each subspace N_α, because $I_\alpha \subseteq I$ for each $\alpha \in A$. Hence, $\vec{c}$ is a nonzero element of the intersection $\bigcap_{\alpha \in A} N_\alpha$. Conversely, suppose that some nonzero vector $\vec{c} = (\gamma_1, \dots, \gamma_n) \in \mathbb{K}^n$ lies in each N_α. Then the equality $\gamma_1 f_1(t) + \cdots + \gamma_n f_n(t) = 0$ holds for each t lying in the union of all intervals I_α, which is a dense subset of I. Thus by continuity, the last equality is valid at each point $t \in I$, which ensures the linear dependence of Φ in the whole interval I. □

Examples 1.2.8: (i) Consider a pair of functions $\{f_1, f_2\}$ defined by

$$f_1(t) = \begin{cases} 0 & \text{if } t < 0 \\ t^2 & \text{if } t \geq 0 \end{cases} \quad \text{and} \quad f_2(t) = \begin{cases} t^2 & \text{if } t < 0 \\ 0 & \text{if } t \geq 0. \end{cases} \tag{1.2.15}$$

These functions which lie in $C^1(\mathbb{R}, \mathbb{R})$ are obviously linearly independent in each interval $[-\varepsilon, \varepsilon]$, $\varepsilon > 0$, because the equality $\gamma_1 f_1(t) + \gamma_2 f_2(t) = 0$ can be valid for some $t < 0$ (or some $t > 0$) only if $\gamma_2 = 0$ (or $\gamma_1 = 0$, respectively). On the other hand, their Wronskian

$$w[f_1, f_2](t) = \begin{vmatrix} f_1(t) & f_2(t) \\ f_1'(t) & f_2'(t) \end{vmatrix} = \begin{cases} \begin{vmatrix} 0 & t^2 \\ 0 & 2t \end{vmatrix} & \text{if } t < 0 \\ \begin{vmatrix} t^2 & 0 \\ 2t & 0 \end{vmatrix} & \text{if } t \geq 0 \end{cases}$$

vanishes at each point $t \in \mathbb{R}$. Thus, each point of the real line is a critical point of $\{f_1, f_2\}$. Moreover, the first row of the above Wronskian shows that $\operatorname{ord} t = 1$ for each $t \neq 0$ and $\operatorname{ord} 0 = 0$. Hence, $I_1 = (-\infty, 0)$ and $I_2 = (0, \infty)$ are intervals of dependence of the pair (1.2.15). One can easily see that the corresponding subspaces (1.2.14) are of form

$$N_1 = \{\vec{c} = (\gamma, 0) \mid \gamma \in \mathbb{R}\} \quad \text{and} \quad N_2 = \{\vec{c} = (0, \gamma) \mid \gamma \in \mathbb{R}\},$$

and thus $N_1 \cap N_2$ is the zero subspace of $\mathbb{R}^2$.

(ii) *Motion in a central force field.* Following Wolsson [W 1], we illustrate the theory of critical points by the motion of a particle under the action of a central force field. Let

$$\vec{r} = \vec{r}(t) = \{x(t), y(t), z(t)\}$$

be the position vector of a particle in a three-dimensional space $\mathbb{R}^3$, which depends on the time t. We will assume that the trajectory is determined by the law

$$\frac{d^2\vec{r}}{dt^2} = a(\vec{r}, t) \cdot \vec{r}, \tag{1.2.16}$$

in which $a = a(x, y, z, t)$ is a given *scalar-valued* function (not necessarilly linear in x, y and z). Consider any solution $\vec{r}(t)$ of (1.2.16), defined on an interval $I = [t_0, T)$. Since the vector equation (1.2.16) is equivalent to the following system of three scalar equations

$$\ddot{x} = a(x, y, z, t)x, \quad \ddot{y} = a(x, y, z, t)y, \quad \ddot{z} = a(x, y, z, t)$$

the third row of the Wronski matrix

$$W(x, y, z) = \begin{pmatrix} x & y & z \\ \dot{x} & \dot{y} & \dot{z} \\ \ddot{x} & \ddot{y} & \ddot{z} \end{pmatrix}$$

is a multiple of the first one. Thus, each $t \in I$ is a critical point of the triple of functions $\{x, y, z\}$. Notice that the sub-Wronskians of order 2 of this triple are equal to the coordinates of the vector product

$$\vec{r} \times \frac{d\vec{r}}{dt} = \begin{vmatrix} \vec{i} & \vec{j} & \vec{k} \\ x & y & z \\ \dot{x} & \dot{y} & \dot{z} \end{vmatrix}$$

where as usual $\vec{i}$, $\vec{j}$ and $\vec{k}$ denote the triple of orthonormal vectors lying in the coordinate axes x, y and z, respectively. An easy computation based on (1.2.16) shows that this vector product does not vary during the motion:

$$\frac{d}{dt}\left(\vec{r} \times \frac{d\vec{r}}{dt}\right) = \frac{d\vec{r}}{dt} \times \frac{d\vec{r}}{dt} + \vec{r} \times \frac{d^2\vec{r}}{dt^2} = \vec{0} + a \cdot \vec{r} \times \vec{r} = \vec{0}.$$

Thus, there exists a constant vector $\vec{c}$ such that

$$\vec{r}(t) \times \frac{d\vec{r}}{dt}(t) = \vec{c} \quad \text{for each } t \in I. \tag{1.2.17}$$

(The vector $\vec{c}$ can be computed as the vector product of the initial position $\vec{r}(t_0)$ and the initial velocity $\frac{d\vec{r}}{dt}(t_0)$). Let us distinguish two cases according to whether the vector $\vec{c}$ is zero or not:

Case 1: If $\vec{c} \neq \vec{0}$, then $\operatorname{ord} t = 2$ for each $t \in I$. In view of Theorem 1.2.5, the triple $\{x, y, z\}$ is linearly dependent in the whole interval I: There exists a nonzero vector $(\alpha, \beta, \gamma) \in \mathbb{R}^3$ such that

$$\alpha \cdot x(t) + \beta \cdot y(t) + \gamma \cdot z(t) = 0 \quad \text{for each } t \in I. \tag{1.2.18}$$

Since the vector $\vec{c}$, equal to the product $\vec{r} \times \frac{d\vec{r}}{dt}$, is orthogonal to each factor of this product, we may choose $(\alpha, \beta, \gamma) = \vec{c}$ in (1.2.18). Consequently, the law (1.2.18) can be written in the form of inner product

$$\langle \vec{r}(t), \vec{c} \rangle = 0.$$

Since (1.2.17) implies now that $\vec{r}(t) \neq \vec{0}$, we have proved the following assertion:

In the case when $\vec{c} \neq \vec{0}$, the whole trajectory of the particle lies in a plane passing through the origin and the particle does not touch the origin at any time (for which the derivative of $\vec{r}$ is finite).

Case 2: If $\vec{c} = \vec{0}$, then either $\operatorname{ord} t = 0$ or $\operatorname{ord} t = 1$, according to whether the vector $\vec{r}(t)$ is zero or not. Let (t_1, t_2) be any (maximal) open subinterval of I on which $\vec{r}(t) \neq \vec{0}$. In view of Theorem 1.2.5, the linear space spanned by the triple

(x, y, z) on (t_1, t_2) is of dimension 1. Consequently, if t goes through (t_1, t_2), the trajectory of the particle lies on a half-line passing through the origin. Wolsson [W 1] gave an example of the system (1.2.16) showing that this half-line can be "turned" whenever the particle passes through the origin (and thus the planarity of the whole trajectory cannot be inferred). This example is posssible only when the uniqueness of the initial-value problem for (1.2.16) fails at $\vec{r} = \vec{0}$.

We have learned that the linear dependence of a given n-tuple of smooth functions in one real variable can be completely investigated by testing all the $2^n - 1$ sub-Wronskians of this n-tuple. Let us finish this section by showing that a mild modification of the original Wronskian (see (1.2.20)) enables us to state a criterion for the linear dependence referring to the only one Wronski-like determinant. However, this result is applicable only in the case that at least one critical point of the given n-tuple is of order $n-1$.

Theorem 1.2.9 [Ši 5]: *Let $f_1, \ldots, f_n \in C^{n-1}(I, \mathbb{K})$, where I is an interval of reals, and let the point $t_0 \in I$ be chosen so that the rank of the $(n-1) \times n$ matrix*

$$\begin{pmatrix} f_1(t_0) & f_2(t_0) & \ldots & f_n(t_0) \\ f_1'(t_0) & f_2'(t_0) & \ldots & f_n'(t_0) \\ \vdots & \vdots & \ddots & \vdots \\ f_1^{(n-2)}(t_0) & f_2^{(n-2)}(t_0) & \ldots & f_n^{(n-2)}(t_0) \end{pmatrix} \tag{1.2.19}$$

is equal to $n-1$. Then the functions $f_1, \ldots, f_n$ are linearly dependent in the interval I if and only if the following determinant

$$\begin{vmatrix} f_1(t_0) & f_2(t_0) & \ldots & f_n(t_0) \\ f_1'(t_0) & f_2'(t_0) & \ldots & f_n'(t_0) \\ \vdots & \vdots & \ddots & \vdots \\ f_1^{(n-2)}(t_0) & f_2^{(n-2)}(t_0) & \ldots & f_n^{(n-2)}(t_0) \\ f_1^{(n-1)}(t) & f_2^{(n-1)}(t) & \ldots & f_n^{(n-1)}(t) \end{vmatrix} \tag{1.2.20}$$

vanishes for each $t \in I$.

Proof: (i) If a nonzero vector $\vec{c} = (\gamma_1, \ldots, \gamma_n) \in \mathbb{K}^n$ satisfies

$$\gamma_1 f_1(t) + \gamma_2 f_2(t) + \cdots + \gamma_n f_n(t) = 0 \quad (t \in I),$$

then $\vec{c}$ is orthogonal to each row of the determinant (1.2.20). Thus, these n rows are linearly dependent and the value of (1.2.20) is zero, for each $t \in I$.

(ii) If the rank of (1.2.19) is $n - 1$, there is no loss of generality in assuming that the first $(n - 1)$ columns in (1.2.19) are linearly independent (renumber f_i if necessary). Supposing the determinant (1.2.20) to be zero, we therefore conclude that the last column of (1.2.20) is (the unique) linear combination of the previous ones: There exist coefficients $\alpha_k = \alpha_k(t)$ such that

$$f_n^{(j)}(t_0) = \sum_{k=1}^{n-1} \alpha_k(t) f_k^{(j)}(t_0) \quad (j=0,1,\ldots,n-2) \tag{1.2.21}$$

and

$$f_n^{(n-1)}(t) = \sum_{k=1}^{n-1} \alpha_k(t) f_k^{(n-1)}(t), \tag{1.2.22}$$

for each $t \in I$. In fact, the coefficients α_k do not depend on the variable t, because they can be uniquely computed from (1.2.21) by using Cramer's rule. Let us introduce a "difference" function

$$\mu(t) := f_n(t) - \sum_{k=1}^{n-1} \alpha_k f_k(t).$$

Since α_k are constants, we easily obtain from (1.2.21) and (1.2.22) that the function μ is a solution of the following initial-value problem

$$x^{(n-1)} = 0 \quad \text{and} \quad x^{(j)}(t_0) = 0 \quad (j=0,1,\ldots,n-2)$$

on the interval I. However, it is well known that this simple problem has a unique solution (identically equal to zero). Hence, μ vanishes on I, which means that f_n is a linear combination of $f_1, \ldots, f_{n-1}$ in I. $\square$

1.3. The Gram determinant

Let (X, μ) be a measure space with σ-additive measure μ and let $L^2(X)$ denote the linear space of all measurable functions $f : X \to \mathbb{K}$ satisfying $\int_X |f|^2 \, d\mu < \infty$. More precisely, the elements of $L^2(X)$ are equivalence classes (i.e. equal almost everywhere) of functions. Thus, an n-tuple $f_1, \ldots, f_n \in L^2(X)$ is linearly dependent (in the set X) if and only if the equality in (1.1.3) holds for almost all $x \in X$. Since $L^2(X)$ is a Hilbert space with the inner product

$$\langle f_1, f_2 \rangle = \int_X f_1 \bar{f}_2 \, d\mu$$

(where $\bar{\ }$ denotes *complex conjugation*, which can be omitted if $\mathbb{K} = \mathbb{R}$), we can define the *Gram matrix* of a given n-tuple of elements $f_1, f_2, \ldots, f_n \in L^2(X)$ as the following matrix of size $n \times n$

$$\Gamma(f_1, f_2, \dots, f_n) := \begin{pmatrix} \langle f_1, f_1\rangle & \langle f_1, f_2\rangle & \dots & \langle f_1, f_n\rangle \\ \langle f_2, f_1\rangle & \langle f_2, f_2\rangle & \dots & \langle f_2, f_n\rangle \\ \vdots & \vdots & \ddots & \vdots \\ \langle f_n, f_1\rangle & \langle f_n, f_2\rangle & \dots & \langle f_n, f_n\rangle \end{pmatrix}. \tag{1.3.1}$$

The value $\gamma = \det \Gamma(f_1, f_2, \dots, f_n)$ is then called the *Gram determinant* (or the Gramian[2]) of the n-tuple $f_1, \dots, f_n$ (cf. [La]). Notice that the Gram determinant is a *constant* while the Wronski determinant is a *function*.

Theorem 1.3.1: *A system of functions $f_1, \dots, f_n \in L^2(X)$ is linearly dependent in the set X if and only if the Gram determinant of $f_1, \dots, f_n$ is equal to zero.*

Proof: Suppose that the functions $f_1, \dots, f_n \in L^2(X)$ are linearly dependent in the set X. Then there exist constants $\alpha_1, \dots, \alpha_n \in \mathbb{K}$, not all 0, such that

$$\alpha_1 f_1(x) + \alpha_2 f_2(x) + \dots + \alpha_n f_n(x) = 0 \qquad \text{for almost all } x \in X.$$

Multiplying this equality by $\bar{f}_j(x)$ and integrating the result over the set X, we get the system of n equalities

$$\langle f_1, f_j\rangle\alpha_1 + \langle f_2, f_j\rangle\alpha_2 + \dots + \langle f_n, f_j\rangle\alpha_n = 0 \quad (1 \le j \le n). \tag{1.3.2}$$

Since this system is linear in $\alpha_1, \dots, \alpha_n$ and has a nontrivial solution, its determinant (equal to the Gram determinant of $f_1, \dots, f_n$) is zero. Conversely, if the Gram determinant of some functions $f_1, \dots, f_n$ is zero, then the system of n equations (1.3.2) has a nontrivial solution $\alpha_1, \dots, \alpha_n$. Multiplying the equation

$$\left\langle \sum_{i=1}^{n} \alpha_i f_i, f_j \right\rangle = 0$$

by the constant $\bar{\alpha_j}$ and adding the results, for $j = 1, 2, \dots, n$, we get

$$\sum_{j=1}^{n} \bar{\alpha_j} \left\langle \sum_{i=1}^{n} \alpha_i f_i, f_j \right\rangle = \left\langle \sum_{i=1}^{n} \alpha_i f_i, \sum_{j=1}^{n} \alpha_j f_j \right\rangle = 0.$$

Thus, $\sum_{i=1}^{n} \alpha_i f_i$ is the zero element of $L^2(X)$, which proves the linear dependence of $f_1, \dots, f_n$. □

[2]After the Danish mathematician J. P. Gram (1850–1916).

1.4. The Casorati determinant

The preceding Theorems 1.2.1 and 1.3.1 that provide criteria of the linear dependence of a finite set of functions $f_i : X \to \mathbb{K}$ $(1 \le i \le n)$ are applicable only in the cases when the functions f_i satisfy some regularity condition (smoothness or integrability). If this is not the case (for example, if the domain of definition X is an nonempty set without any structure like topology or measure), we will introduce the matrix of size $n \times n$

$$C(f_1, f_2, \dots, f_n; \boldsymbol{x}) := \begin{pmatrix} f_1(x_1) & f_2(x_1) & \dots & f_n(x_1) \\ f_1(x_2) & f_2(x_2) & \dots & f_n(x_2) \\ \vdots & \vdots & \ddots & \vdots \\ f_1(x_n) & f_2(x_n) & \dots & f_n(x_n) \end{pmatrix} \tag{1.4.1}$$

for each n-tuple $\boldsymbol{x} = (x_1, x_2, \dots, x_n) \in X^n$. Following standard terminology (cf. [AD] or [Be]), we call (1.4.1) the *Casorati matrix* of the n-tuple of functions $f_1, \dots, f_n$ taken at the point $\boldsymbol{x} \in X^n$. The value of $\det C(f_1, \dots, f_n; \boldsymbol{x})$, called the *Casorati determinant*[3] of $f_1, \dots, f_n$, is a mapping $\det C : X^n \to \mathbb{K}$. In fact, the origin of this determinant is associated with the solution of a linear difference equation of the form

$$f(x+N) + p_1(x) f(x+N-1) + \dots + p_N(x) f(x) = 0 \quad (x = 0, 1, 2, \dots)$$

in which p_i are given coefficient functions, while f is an unknown. Let us mention (without proof) the fact that if $p_N(x) \neq 0$ for each x and if $f_1, \dots, f_n$ is any n-tuple of solutions of the above difference equation, then the so-called Casorati determinant $\det C(f_1, f_2, \dots, f_N; x)$, defined as

$$\begin{vmatrix} f_1(x) & f_2(x) & \dots & f_N(x) \\ f_1(x+1) & f_2(x+1) & \dots & f_N(x+1) \\ \vdots & \vdots & \ddots & \vdots \\ f_1(x+N-1) & f_2(x+N-1) & \dots & f_N(x+N-1) \end{vmatrix},$$

vanishes (in the variable x) either anywhere or nowhere, according to whether the n-tuple of functions $f_1, \dots, f_n$ is linearly dependent or not.

The basic property of the Casorati determinants is described in the following theorem. For some other main properties see the classical book of Montel [M].

Theorem 1.4.1: *A system of functions $f_i : X \to \mathbb{K}$ $(i = 1, 2, \dots, n)$ is linearly dependent in the set X if and only if the Casorati determinant of $f_1, \dots, f_n$ is equal to zero at each point of X^n.*

---

[3] After the Italian mathematician Felice Casorati (1835–1890).

Proof: If the functions $f_1, \ldots, f_n$ are linearly dependent in the set X, then there exist constants $\alpha_1, \ldots, \alpha_n \in \mathbb{K}$, not all zeroes, such that

$$\alpha_1 f_1(x) + \alpha_2 f_2(x) + \cdots + \alpha_n f_n(x) = 0 \qquad (x \in X).$$

Setting here $x = x_i$, where $x_1, \ldots, x_n \in X$ are arbitrary elements, we get a system of n equalities, which are linear in $\alpha_1, \ldots, \alpha_n$. Since this system has a nontrivial solution, its determinant (equal to the Casorati determinant of $f_1, \ldots, f_n$ at the point $(x_1, \ldots, x_n)$) is equal to zero. Conversely, suppose that the Casorati determinant of $f_1, \ldots, f_n$ vanishes at each point of X^n. We restrict our attention to the case when $f_1(x_0) \neq 0$ for some $x_0 \in X$ (otherwise $f_1, \ldots, f_n$ are clearly linearly dependent in the set X). Then there exists the smallest integer p $(1 < p \leq n)$ so that the Casorati determinant of $f_1, \ldots, f_p$ vanishes at each point of X^p. In view of this definition, there exist elements $u_1, \ldots, u_{p-1} \in X$ satisfying

$$\begin{vmatrix} f_1(u_1) & f_2(u_1) & \ldots & f_{p-1}(u_1) \\ f_1(u_2) & f_2(u_2) & \ldots & f_{p-1}(u_2) \\ \vdots & \vdots & \ddots & \vdots \\ f_1(u_{p-1}) & f_2(u_{p-1}) & \ldots & f_{p-1}(u_{p-1}) \end{vmatrix} \neq 0, \tag{1.4.3}$$

while the equality

$$\begin{vmatrix} f_1(u_1) & f_2(u_1) & \ldots & f_{p-1}(u_1) & f_p(u_1) \\ f_1(u_2) & f_2(u_2) & \ldots & f_{p-1}(u_2) & f_p(u_2) \\ \vdots & \vdots & \ddots & \vdots & \vdots \\ f_1(u_{p-1}) & f_2(u_{p-1}) & \ldots & f_{p-1}(u_{p-1}) & f_p(u_{p-1}) \\ f_1(x) & f_2(x) & \ldots & f_{p-1}(x) & f_p(x) \end{vmatrix} = 0 \tag{1.4.4}$$

holds for each $x \in X$. (Notice that the left-hand side of (1.4.4) is the value of the Casorati determinant of $f_1, \ldots, f_p$ at the point $(u_1, \ldots, u_{p-1}, x)$). Developing the determinant in (1.4.4) with respect to the last row and taking in account (1.4.3), we conclude that the function f_p is a linear combination of $f_1, \ldots, f_{p-1}$. This proves the linear dependence of $f_1, \ldots, f_n$ in the set X. □

Now we characterize functions $F \colon X^n \to \mathbb{K}$ that admit the representation in the form of some Casorati determinant

$$F(x_1, x_2, \ldots, x_n) = \det \begin{pmatrix} f_1(x_1) & f_2(x_1) & \ldots & f_n(x_1) \\ f_1(x_2) & f_2(x_2) & \ldots & f_n(x_2) \\ \vdots & \vdots & \ddots & \vdots \\ f_1(x_n) & f_2(x_n) & \ldots & f_n(x_n) \end{pmatrix} \tag{1.4.5}$$

of a suitable n-tuple of functions $f_i \colon X \to \mathbb{K}$, $1 \leq i \leq n$. Note firstly that the well known rule concerning the permutations of rows in a determinant yields immediately a necessary condition of the existence of (1.4.5): the function F must be *antisymmetric* in the variables $x_1, x_2, \ldots, x_n$. However, this condition is not sufficient in general (see Example 1.4.4 below). Let us add that the representation (1.4.5) is "useful" only in the case when the functions $f_1, \ldots, f_n$ are linearly independent in the set X. Thus, we will assume that the function F under consideration has at least one nonzero value in X^n.

Theorem 1.4.2 [Ši 1]: *Let $F\colon X \to \mathbb{K}$ be a function such that $F(\boldsymbol{u}) \neq 0$ for some $\boldsymbol{u} = (u_1, u_2, \ldots, u_n) \in X^n$. Choose such a $\mathbf{u}$, denote $c = F(\boldsymbol{u})$ and introduce the partial functions $F_i\colon X \to \mathbb{K}$ by*

$$F_i(x) = F(u_1, \ldots, u_{i-1}, x, u_{i+1}, \ldots, u_n) \qquad (x \in X,\ 1 \le i \le n). \qquad (1.4.6)$$

Then F has a representation (1.4.5) *on X^n if and only if the equality*

$$F(\boldsymbol{x}) = \frac{1}{c^{n-1}} \cdot \det \begin{pmatrix} F_1(x_1) & F_2(x_1) & \ldots & F_n(x_1) \\ F_1(x_2) & F_2(x_2) & \ldots & F_n(x_2) \\ \vdots & \vdots & \ddots & \vdots \\ F_1(x_n) & F_2(x_n) & \ldots & F_n(x_n) \end{pmatrix} \qquad (1.4.7)$$

holds for each $\boldsymbol{x} = (x_1, x_2, \ldots, x_n) \in X^n$. Moreover, each function f_i from any representation (1.4.5) *is then a linear combination of the partial functions $F_1, \ldots, F_n$.*

Proof: We need to show that the function F of the form (1.4.5) satisfies (1.4.7) for each $\boldsymbol{x} = (x_1, x_2, \ldots, x_n) \in X^n$, because the converse conclusion is obvious: if (1.4.7) holds , then F is of type (1.4.5), for instance with $f_i = F_i$ $(1 \le i \le n-1)$ and $f_n = c^{1-n} \cdot F_n$. So let F be as in (1.4.5), with suitable functions $f_1, \ldots, f_n$. Developing the determinant in the right-hand side of (1.4.5) with respect to the first row and using the notation $x = x_1 \in X$ and $\boldsymbol{y} = (x_2, x_3, \ldots, x_n) \in X^{n-1}$, we obtain the representation

$$F(\boldsymbol{x}) = F(x, \boldsymbol{y}) = \sum_{j=1}^{n} f_j(x) g_j(\boldsymbol{y})$$

with some functions $g_j\colon X^{n-1} \to \mathbb{K}$. Setting here n values

$$\boldsymbol{y} = \boldsymbol{y}^i = (u_1, \ldots, u_{i-1}, u_{i+1}, \ldots, u_n) \quad (\text{for } i = 1, 2, \ldots, n)$$

where $u_1, u_2, \ldots, u_n$ are the components of the n-tuple $\boldsymbol{u} \in X^n$ mentioned in the statement of the theorem, and taking in account that $F(x, \boldsymbol{y}^i) = (-1)^{i-1} F_i(x)$ (see (1.4.6)), we get the following system of n equalities

$$(-1)^{i-1} F_i(x) = \sum_{j=1}^{n} f_j(x) g_j(\boldsymbol{y}^i) \quad (x \in X,\ 1 \le i \le n). \qquad (1.4.8)$$

The latter is a linear system in $f_1(x), \dots, f_n(x)$. Its matrix $[g_j(\boldsymbol{y}^i)]_{i,j=1}^n$ is independent in x and nonsingular, because of

$$\det[g_j(\boldsymbol{y}^i)]_{ij} \cdot \det[f_j(u_k)]_{jk} = \det\left[\sum_{j=1}^n f_j(u_k) g_j(\boldsymbol{y}^i)\right]_{ik}$$
$$= \det[(-1)^{i-1} F_i(u_k)]_{ik} = \pm c^n \neq 0\,.$$

Here we have used the fact that the partial functions F_i clearly satisfy the condition

$$F_i(u_k) = \begin{cases} c & (i = k) \\ 0 & (i \neq k). \end{cases} \tag{1.4.9}$$

Consequently, the values $f_1(x), \dots, f_n(x)$ are uniquely determined from the linear system (1.4.8) by means of Cramer's rule. Thus, there exist constants $\alpha_{ik} \in \mathbb{K}$ such that

$$f_i(x) = \sum_{k=1}^n \alpha_{ik} F_k(x) \qquad (x \in X,\, 1 \leq i \leq n) \tag{1.4.10}$$

(which proves the last conclusion of Theorem 1.4.2). From (1.4.5) and (1.4.10) we obtain the representation

$$F(\boldsymbol{x}) = \det A \cdot \det \begin{pmatrix} F_1(x_1) & F_2(x_1) & \dots & F_n(x_1) \\ F_1(x_2) & F_2(x_2) & \dots & F_n(x_2) \\ \vdots & \vdots & \ddots & \vdots \\ F_1(x_n) & F_2(x_n) & \dots & F_n(x_n) \end{pmatrix} \tag{1.4.11}$$

where A is the constant $n \times n$ matrix with entries α_{ik}. It remains to verify that $\det A = c^{1-n}$. Setting $\boldsymbol{x} = \boldsymbol{u}$ in (1.4.11) and using (1.4.9), we find that

$$c = F(\boldsymbol{u}) = \det A \cdot \det(cE_n) = \det A \cdot c^n$$

where E_n is the unit matrix of order n. Thus, we have $\det A = c^{1-n}$, and the proof is complete. □

Theorem 1.4.2 implies that each function F equal to the Casorati determinant of some linearly independent n-tuple of functions is uniquely determined by the n-tuple of the partial functions F_i from (1.4.6), taken at a chosen n-tuple $\boldsymbol{u}$. The following lemma shows that the condition (1.4.9) is the unique general property of such n-tuples of the partial functions.

Lemma 1.4.3: *Let $F_i \colon X \to \mathbb{K}$ $(1 \le i \le n)$ be any n-tuple of functions satisfying* (1.4.9), *where $c \in \mathbb{K}$ is a nonzero constant and $u_1, u_2, \ldots, u_n$ are chosen elements of X. Then the function $F \colon X \to \mathbb{K}$ defined by* (1.4.7) *satisfies* (1.4.6).

Proof: If F is as in (1.4.7), then (1.4.9) implies that

$$\begin{aligned} F(x, u_2, u_3, \ldots, u_n) &= \frac{1}{c^{n-1}} \cdot \det \begin{pmatrix} F_1(x) & F_2(x) & F_3(x) & \ldots & F_n(x) \\ 0 & c & 0 & \ldots & 0 \\ 0 & 0 & c & \ldots & 0 \\ \vdots & \vdots & \vdots & \ddots & \vdots \\ 0 & 0 & 0 & \ldots & c \end{pmatrix} \\ &= \frac{1}{c^{n-1}} \cdot F_1(x) \cdot c^{n-1} = F_1(x), \end{aligned}$$

which proves (1.4.6) with $i = 1$. The proof of the other equalitites in (1.4.6) is analogous. □

Example 1.4.4: We find all odd integers $k \ge 1$ for which the polynomial

$$F(x_1, x_2) = (x_1 - x_2)^k$$

(in two real variables x_1, x_2) is the Casorati determinant of some pair of real-valued functions. Since $F(1,0) = 1 \ne 0$, Theorem 1.4.2 implies that F has the mentioned property if and only if the partial functions

$$F_1(x) = F(x, 0) = x^k \quad \text{and} \quad F_2(x) = F(1, x) = (1 - x)^k$$

satisfy the identity

$$F(x_1, x_2) = \det \begin{pmatrix} F_1(x_1) & F_2(x_1) \\ F_1(x_2) & F_2(x_2) \end{pmatrix}$$

or, equivalently,

$$(x_1 - x_2)^k = F_1(x_1)F_2(x_2) - F_1(x_2)F_2(x_1) = x_1^k(1 - x_2)^k - x_2^k(1 - x_1)^k .$$

Equating the coefficients of power x_1^k on both sides of the last equality, we get the condition

$$1 = (1 - x_2)^k - (-1)^k x_2^k ,$$

which clearly holds only if $k = 1$. Conversely, the polynomial $F(x_1, x_2) = x_1 - x_2$ is the Casorati determinant of the pair $f_1(x) = x$ and $f_2(x) = 1$, $x \in \mathbb{R}$.

Let us finish this chapter by stating an important consequence of Theorem 1.4.2 saying that two n-tuples of functions with the common domain of definition are two bases of the same linear space of dimension n if and only if their Casorati determinants are the same mappings – up to a multiplicative nonzero constant.

Corollary 1.4.5: *Let* $f_i : X \to \mathbb{K}$ *and* $g_i : X \to \mathbb{K}$ $(1 \le i \le n)$ *be two* n*-tuples of functions and let* $F : X^n \to \mathbb{K}$ *and* $G : X^n \to \mathbb{K}$ *be their Casorati determinants, respectively. Suppose that each of the* n*-tuples* $\{f_i\}_{i=1}^n$ *and* $\{g_i\}_{i=1}^n$ *is linearly independent in the set* X*. Then these two* n*-tuples are bases of the same linear functional space (over the field* $\mathbb{K}$*) if and only if there exists a nonzero constant* $\beta \in \mathbb{K}$ *such that the equality* $F(\boldsymbol{x}) = \beta \cdot G(\boldsymbol{x})$ *holds for each* $\boldsymbol{x} \in X^n$*.*

Proof: If $\{f_i\}_{i=1}^n$ and $\{g_i\}_{i=1}^n$ are two bases of the same linear space, then there exists a constant nonsingular $n \times n$ matrix $A = [\alpha_{ij}]_{i,j=1}^n$ such that

$$\beta := \det A \neq 0 \quad \text{and} \quad f_i(x) = \sum_{j=1}^{n} \alpha_{ij} g_j(x) \quad \text{for each } x \in X.$$

The Casorati determinant F for this representation of f_i becomes

$$\begin{aligned} F(\boldsymbol{x}) = \det[f_i(x_k)]_{ik} &= \det\big([\alpha_{ij}]_{ij} \cdot [g_j(x_k)]_{jk}\big) = \det A \cdot \det[g_j(x_k)]_{jk} \\ &= \beta \cdot G(\boldsymbol{x}) \quad (\boldsymbol{x} = (x_1, x_2, \dots, x_n) \in X^n). \end{aligned}$$

Conversely, suppose that $F(\boldsymbol{x}) = \beta \cdot G(\boldsymbol{x})$, for some constant $\beta \neq 0$ and for each $\boldsymbol{x} \in X^n$. Since the n-tuple $\{f_i\}_{i=1}^n$ is assumed to be linearly independent in the set X, Theorem 1.4.1 implies that $F(\boldsymbol{u}) \neq 0$ for some $\boldsymbol{u} \in X^n$. Then, of course $G(\boldsymbol{u}) \neq 0$ and the last sentence of Theorem 1.4.2 implies that for each i, both f_i and g_i are linear combinations of the same n-tuple of the partial functions $F_1, \dots, F_n$ defined by means of (1.4.6). Consequently, $\{f_i\}_{i=1}^n$, $\{g_i\}_{i=1}^n$ and $\{F_i\}_{i=1}^n$ are three bases of the same linear space. □

2 BASIC DECOMPOSITION THEOREMS FOR FUNCTIONS OF TWO VARIABLES

In this chapter, we solve the fundamental problem of finding necessary and sufficient conditions for a given scalar function h of two variables (say x and y) to be decomposed into the form

$$h(x, y) = \sum_{i=1}^{n} f_i(x) g_i(y) \tag{2a}$$

with a prescribed integer $n \geq 1$. We conclude with functional and differential equations for matrix-valued functions H permitting a factorization

$$H(x, y) = F(x) \cdot G(y) \tag{2b}$$

where $\cdot$ stands for the usual matrix multiplication.

2.1. The Wronski determinant of a function of two variables

The Wronski matrix (1.2.1) of an n-*tuple* of smooth functions f_j in *one* scalar variable is composed in such a way that the index of the row determines the order of the derivative (diminished by 1) of the functions f_j, while the index j of the column determines the arrangement of the particular functions f_j. For a *single* smooth function h of *two* scalar variables x, y, we may consider an analogous matrix built from the partial derivatives

$$W_n h := \begin{pmatrix} h & h_y & \dots & h_{y^{n-1}} \\ h_x & h_{xy} & \dots & h_{xy^{n-1}} \\ \vdots & \vdots & \ddots & \vdots \\ h_{x^{n-1}} & h_{x^{n-1}y} & \dots & h_{x^{n-1}y^{n-1}} \end{pmatrix} \qquad (n = 1, 2, \dots). \qquad (2.1.1)$$

In accordance with Gauchman and Rubel [GR], we call (2.1.1) the *Wronski matrix of the n-th order* of the function h. The value of its determinant $\det W_n h$ is called the *Wronski determinant* (or also the *Wronskian*) of the n-th order of the function h. To the best of our knowledge, the first appearance of such a kind of matrix is due to Cyparissos Stéphanos [St 1-2] in 1904 in the Section of *Arithmetics and Algebra* at the Third International Congress of Mathematicians in Heidelberg. Notice that, in terms of Chapter 1, $\det W_n h$ is the Wronskian of the n-tuple of the functions $h, h_y, \dots, h_{y^{n-1}}$ in the variable x and, at the same time, it is the Wronskian of the n-tuple of the functions $h, h_x, \dots, h_{x^{n-1}}$ in the variable y.

The crucial meaning of the Wronski matrices (2.1.1) in solving the decomposition problem (2a) is described in the following theorem. Since we have already mentioned the history of the Wronski matrices (2.1.1) in the Prologue, we are now ready to formulate the following result of Neuman (1981).

Theorem 2.1.1 [N 1]: *Let I and J be two intervals in $\mathbb{R}$ and let $n \geq 1$ be an integer. Suppose that a function $h\colon I \times J \to \mathbb{K}$ has the partial derivative $h_{x^n y^n}$ continuous at each point of the rectangle $I \times J$. If h is of the form* (2a) *on $I \times J$, then*

$$\det W_{n+1} h(x, y) = 0 \quad \textit{at each point } (x, y) \in I \times J. \qquad (2.1.2)$$

Conversely, if h satisfies (2.1.2) *and if in addition*

$$\det W_n h(x, y) \neq 0 \quad \textit{at each point } (x, y) \in I \times J, \qquad (2.1.3)$$

then h has a decomposition (2a) *on $I \times J$, with linearly independent components $f_i \in C^n(I)$ and $g_i \in C^n(J)$, $1 \leq i \leq n$.*

Proof: (i) If h is of the form (2a) and if the value of $y \in J$ is fixed, then each of the functions

$$h(-, y),\ h_y(-, y),\ \dots,\ h_{y^n}(-, y) \qquad (2.1.4)$$

is clearly a linear combination of n functions $f_1, \dots, f_n$ in the interval I. (Notice that (2.1.4) is an example of (1.1.2) with operators $L_i g = \frac{d^{i-1}}{dy^{i-1}} g(y_i)$, $1 \leq i \leq n+1$.) Thus the $(n+1)$ functions (2.1.4) are linearly dependent in I and Theorem 1.2.1 implies that their Wronskian (equal to $\det W_n h(-, y)$) vanishes at each point of I. Thus (2.1.2) holds.

(ii) If h satisfies (2.1.2) and (2.1.3), then the second part of Theorem 1.2.1 implies that the function $h_{y^n}(-,y)$ is a linear combination of the n-tuple

$$h(-,y),\ h_y(-,y),\ \ldots,\ h_{y^{n-1}}(-,y)$$

in the interval I, for each (fixed) $y \in J$. Hence, there exist n coefficients $\alpha_i : J \to \mathbb{K}$ such that

$$h_{y^n}(x,y) = \sum_{j=1}^{n} \alpha_j(y) h_{y^{j-1}}(x,y) \quad (x \in I,\ y \in J). \tag{2.1.5}$$

These α_j are continuous on J, because they can be computed from the linear system

$$h_{x^i y^n}(x,y) = \sum_{j=1}^{n} \alpha_j(y) h_{x^i y^{j-1}}(x,y) \quad (i=0,1,\ldots,n-1)$$

by using Cramer's rule. In view of (2.1.5), the function $h(x,-)$ is a solution of the differential equation

$$\frac{d^n g}{dy^n} - \alpha_{n-1}(y)\frac{d^{n-1}g}{dy^{n-1}} - \ldots - \alpha_0(y)g = 0 \quad (g = g(y),\ y \in J), \tag{2.1.6}$$

which does not depend on $x \in I$. Since the coefficients α_j are continuous, equation (2.1.6) has a fundamental set of solutions $g_1, \ldots, g_n$ lying in $C^n(J)$. Now (2.1.5) implies that the function $h(x,-)$ is a linear combination of $g_1, \ldots, g_n$ in the set J. Hence, there exist coefficients $f_i : I \to \mathbb{K}$ such that

$$h(x,y) = \sum_{i=1}^{n} f_i(x) g_i(y) \quad (x \in I,\ y \in J).$$

These f_i are of class $C^n(I)$, because they can be computed from the linear system

$$h_{y^j}(x,y) = \sum_{i=1}^{n} f_i(x) g_i^{(j)}(y) \quad (j=0,1,\ldots,n-1)$$

by using Cramer's rule. Finally, an easy computation shows that the equality

$$\det W_n h(x,y) = \det W(f_1,\ldots,f_n;x) \cdot \det W(g_1,\ldots,g_n;y) \tag{2.1.7}$$

holds at each point $(x,y) \in I \times J$. (The determinants in the right-hand side are the Wronskians defined in Section 1.2.) Thus Theorem 1.2.1 and condition (2.1.3) imply that the n-tuples $\{f_i\}_{i=1}^n$ and $\{g_i\}_{i=1}^n$ are linearly independent in the intervals I and J, respectively. $\square$

Remarks 2.1.2: **(i)** In the original statement of Theorem 2.1.1 in [N 1], the question of all possible decompositions (2a) of the given function h under the conditions (2.1.2) and (2.1.3) is solved. We refer the reader to Section 2.3, where this problem will be treated in general.

(ii) Let us explain to what extent the additional condition (2.1.3) from the second part of Theorem 2.1.1 is restrictive. It is clearly no added restriction to assume that the n-tuples of the components $\{f_i\}_{i=1}^n$ and $\{g_i\}_{i=1}^n$ are linearly independent in the intervals I and J, respectively (otherwise the number n of products in (2a) can be reduced as explained in the beginning of Section 2.3 below). In view of the identity (2.1.7) and the part (ii) of Remarks 1.2.2, the restriction (2.1.3) means that the n-tuples $\{f_i\}_{i=1}^n$ and $\{g_i\}_{i=1}^n$ are not only linearly independent, but that they form even fundamental sets of solutions of two linear differential equations of order n, defined on the intervals I and J, respectively.

(iii) Rassias [Ra 1] proved (independently from F. Neuman) a *local existence* version of Theorem 2.1.1: *If a function h satisfies* (2.1.2) *and if the condition* $\det W_n h(x,y) \neq 0$ *holds at some point* $(x,y) \in I \times J$, *then h has a decomposition* (2a) *in some subregion of the rectangle* $I \times J$. This result follows from Theorem 2.1.1 applied to a sufficiently small "subrectangle" of $I \times J$ in which the Wronskian $\det W_n h$ has no zero value.

(iv) In the case when the inequality $\det W_n h(x_0, y_0) \neq 0$ holds at some point $(x_0, y_0) \in I \times J$ (but not everywhere on $I \times J$), a necessary and sufficient condition for h to have a *global* decomposition (2a) is given in Theorem 4.2.5. This by-product of the approximation theory developed in [Ši 3] involves a functional modification (6.2.3) of the original Wronski determinant (2.1.1).

Counter-example 2.1.3 [Ra 1]**:** In the following we will show that the single condition (2.1.2) is not sufficient for a function h to be of the form (2a) *on the whole* rectangle $I \times J$. (In other words, the condition (2.1.3) in Theorem 2.1.1 is *essential.*) To show this, Th. M. Rassias in 1984 considered the function $h(x,y) = xy^2 + y|y|$, which cannot be expressed in the form $h(x,y) = f(x)g(y)$. It is clear that h has the continuous mixed derivative h_{xy} and that

$$\det W_2 h(x,y) = \begin{vmatrix} h(x,y) & h_y(x,y) \\ h_x(x,y) & h_{xy}(x,y) \end{vmatrix} = \begin{vmatrix} xy^2 + y|y| & 2xy + 2|y| \\ y^2 & 2y \end{vmatrix} = 0$$

at each point $(x,y) \in \mathbb{R}^2$. The counter-example has thus been provided. In 1988, H. Gauchman and L. A. Rubel gave another counter-example, which is even of the class C^∞ (see also Remark 8.2.4). In the same paper, they proved the following:

Theorem 2.1.4 [GR]**:** *Let I and J be two intervals in $\mathbb{R}$ and let $n \geq 1$ be an integer. Suppose that a function $h\colon I \times J \to \mathbb{K}$ is analytic on $I \times J$. If there exist (small but nonempty) open subintervals $I_0 \subseteq I$ and $J_0 \subseteq J$ such that*

$$\det W_{n+1} h(x,y) = 0 \quad \textit{at each point } (x,y) \in I_0 \times J_0\,, \tag{2.1.8}$$

then h is of the form (2a) *on $I \times J$, with some components f_i and g_i, analytic on I and J, respectively.*

Proof: Assume that h has at least one nonzero value in $I \times J$ (otherwise the conclusion of the theorem is trivial). Let $n \geq 1$ be the smallest integer such that (2.1.8) holds, for some subintervals $I_0 \subseteq I$ and $J_0 \subseteq J$. Clearly, we have $\det W_n h(x_0, y_0) \neq 0$ at some point $(x_0, y_0) \in I_0 \times J_0$. Moreover, we may assume that

$$\det W_n h(x,y) \neq 0 \quad \text{at each point } (x,y) \in I_0 \times J_0 \,,$$

otherwise $I_0 \times J_0$ can be reduced to a small neighbourhood of the point (x_0, y_0). Then by Theorem 2.1.1, we have a representation

$$h(x,y) = \sum_{i=1}^{n} f_i(x) g_i(y) \qquad (x \in I_0 \,, y \in J_0) \tag{2.1.9}$$

where the n-tuple $f_1, \ldots, f_n$ is linearly independent in I_0. In view of Theorem 1.4.1, the last property ensures the existence of the points $x_1, \ldots, x_n$ in I_0 for which the matrix $[f_i(x_j)]_{i,j=1}^n$ is nonsingular. Setting successively $x = x_1, \ldots, x_n$ in (2.1.9) and using Cramer's rule, we conclude that

$$g_i(y) = \sum_{j=1}^{n} \alpha_{ij} h(x_j, y) \qquad (y \in J_0,\ i = 1, 2, \ldots, n) \tag{2.1.10}$$

with suitable constants $\alpha_{ij} \in \mathbb{K}$. Since the sums in the right-hand side of (2.1.10) are (in the variable y) analytic on the whole interval J, we get from (2.1.10) that the components g_i from the representation (2.1.9) have analytic continuations to J. Similarly, the components f_i from (2.1.9) have analytic continuations to the whole interval I. Consequently, the sum in the right-hand side of (2.1.9) has an analytic continuation to $I \times J$. This continuation is unique, because of the *uniqueness* theorem (for analytic functions in two variables). Hence, the formula (2.1.9) remains to be true on the whole rectangle $I \times J$. □

As an appendix to Theorem 2.1.1, we now prove a representation formula, which enables us to construct the "unknown" decomposition (2a) from the partial derivatives of the function h.

Theorem 2.1.5 [ČŠ 2]: *Under the conditions* (2.1.2) *and* (2.1.3), *the function h can be decomposed on the whole rectangle $I \times J$ into the following product of three* *matrices of sizes* $1 \times n$, $n \times n$ *and* $n \times 1$, *respectively:*

$$h(x,y) = \big(h(x,y_0), h_y(x,y_0), \ldots, h_{y^{n-1}}(x,y_0)\big) \times W_n^{-1}h(x_0,y_0) \cdot \begin{pmatrix} h(x_0,y) \\ h_x(x_0,y) \\ \vdots \\ h_{x^{n-1}}(x_0,y) \end{pmatrix}. \tag{2.1.11}$$

Here $(x_0, y_0) \in I \times J$ *is an arbitrarily chosen point and* $W_n^{-1}h$ *denotes the inverse of the Wronski matrix* $W_n h$.

Proof: Suppose that a given function $h\colon I \times J \to \mathbb{K}$ satisfies the conditions (2.1.2) and (2.1.3). According to Theorem 2.1.1 and part (ii) of Remarks 2.1.2, the function h has a decomposition

$$h(x,y) = \sum_{i=1}^{n} f_i(x)g_i(y) \qquad (x \in I,\ y \in J) \tag{2.1.12}$$

in which $\{f_i\}_{i=1}^n$ and $\{g_i\}_{i=1}^n$ form fundamental sets of solutions of two linear differential equations of order n. Differentiating (2.1.12) with respect to y yields

$$h_{y^j}(x,y_0) = \sum_{i=1}^{n} f_i(x)g_i^{(j)}(y_0) \qquad (x \in I,\ 0 \le j \le n-1) \tag{2.1.13}$$

for each (fixed) $y_0 \in J$. We may consider (2.1.13) to be a linear algebraic system of n equations with "unknowns" $f_1(x), \cdots, f_n(x)$. Since the matrix of this system is equal to the nonsingular Wronski matrix of the n-tuple $g_1, \ldots, g_n$ (taken at the point $y = y_0$), we can apply Cramer's rule to this system and conclude that

$$f_i(x) = \sum_{j=0}^{n-1} \alpha_{ij} h_{y^j}(x,y_0) \qquad (x \in I,\ 1 \le i \le n) \tag{2.1.14}$$

where $\alpha_{ij} \in \mathbb{K}$ are n^2 suitable constants. Analogously, we have

$$g_i(y) = \sum_{j=0}^{n-1} \beta_{ij} h_{x^j}(x_0,y) \qquad (y \in J,\ 1 \le i \le n). \tag{2.1.15}$$

Substituting (2.14) and (2.1.15) into (2.1.12), we obtain the representation

$$h(x,y) = \big(h(x,y_0), h_y(x,y_0), \ldots, h_{y^{n-1}}(x,y_0)\big) \cdot A \times \begin{pmatrix} h(x_0,y) \\ h_x(x_0,y) \\ \vdots \\ h_{x^{n-1}}(x_0,y) \end{pmatrix} \tag{2.1.16}$$

with a constant matrix A of size $(n \times n)$. To finish the proof, it remains to verify that A is the inverse of $W_n h(x_0, y_0)$. This can be done very easily, because differentiating in (2.1.16) yields the matrix identity

$$W_n h(x,y) = W_n h(x,y_0) \cdot A \cdot W_n h(x_0,y) \qquad (x \in I,\ y \in J).$$

Setting here $x = x_0$ and $y = y_0$, we find that $A = W_n^{-1} h(x_0, y_0)$. □

2.2. The Casorati determinant of a function of two variables

The notion of the Casorati matrix has been defined in Section 1.4 for an arbitrary n-*tuple* of functions of *one* variable. Following Neuman [N 1], we now introduce it for *one* function of *two* variables in a way analogous to the procedure for the Wronski matrix in Section 2.1. For any function $h\colon X \times Y \to \mathbb{K}$, where X and Y are arbitrary nonempty sets, F. Neuman has defined

$$C_n h(\boldsymbol{x},\boldsymbol{y}) := \begin{pmatrix} h(x_1,y_1) & h(x_1,y_2) & \ldots & h(x_1,y_n) \\ h(x_2,y_1) & h(x_2,y_2) & \ldots & h(x_2,y_n) \\ \vdots & \vdots & \ddots & \vdots \\ h(x_n,y_1) & h(x_n,y_2) & \ldots & h(x_n,y_n) \end{pmatrix} \tag{2.2.1}$$

for each $\boldsymbol{x} = (x_1, \ldots, x_n) \in X^n$ and each $\boldsymbol{y} = (y_1, \ldots, y_n) \in Y^n$. The matrix $C_n h$ is called the *Casorati matrix of the n-th order* of the function h. The value of the determinant

$$\det C_n h(\boldsymbol{x},\boldsymbol{y}) = \begin{vmatrix} h(x_1,y_1) & h(x_1,y_2) & \ldots & h(x_1,y_n) \\ h(x_2,y_1) & h(x_2,y_2) & \ldots & h(x_2,y_n) \\ \vdots & \vdots & \ddots & \vdots \\ h(x_n,y_1) & h(x_n,y_2) & \ldots & h(x_n,y_n) \end{vmatrix} \tag{2.2.2}$$

is called the *Casorati determinant of n-th order* of the function h. Notice that the Casorati determinant $\det C_n h(\boldsymbol{x},\boldsymbol{y})$ is (in the variable $\boldsymbol{x}$) the Casorati determinant of the n-tuple $h(-, y_1), \ldots, h(-, y_n)$, for each (fixed) $\boldsymbol{y} = (y_1, \ldots, y_n) \in Y^n$, and, at the same time, $\det C_n h(\boldsymbol{x}, \boldsymbol{y})$ is (in the variable $\boldsymbol{y}$) the Casorati determinant of the n-tuple $h(x_1, -), \ldots, h(x_n, -)$, for each (fixed) $\boldsymbol{x} = (x_1, \ldots, x_n) \in X^n$.

Theorem 2.2.1 [N 1]: *Let $h : X \times Y \to \mathbb{K}$ be a function, where X and Y are arbitrary nonempty sets. If h is of the form* (2a) *on $X \times Y$ for some $n \geq 1$, then*

$$\det C_{n+1} h(\boldsymbol{x}, \boldsymbol{y}) = 0 \qquad \textit{for each } \boldsymbol{x} \in X^{n+1} \textit{ and } \boldsymbol{y} \in Y^{n+1}. \tag{2.2.3}$$

Conversely, if h satisfies (2.2.3) *for some $n \geq 1$ and if in addition*

$$\det C_n h(\boldsymbol{x}^0, \boldsymbol{y}^0) \neq 0 \qquad \textit{for some } \boldsymbol{x}^0 \in X^n \textit{ and } \boldsymbol{y}^0 \in Y^n, \tag{2.2.4}$$

then h has a decomposition (2a) *on $X \times Y$ in which the n-tuples of components $\{f_i\}_{i=1}^n$ and $\{g_i\}_{i=1}^n$ are linearly independent in X and Y, respectively. Moreover, the function h can then be decomposed on $X \times Y$ into the following product of three matrices of sizes $1 \times n$, $n \times n$ and $n \times 1$:*

$$h(x, y) = \big(h(x, y_1), h(x, y_2), \ldots, h(x, y_n)\big) \times C_n^{-1} h(\boldsymbol{x}^0, \boldsymbol{y}^0) \cdot \begin{pmatrix} h(x_1, y) \\ h(x_2, y) \\ \vdots \\ h(x_n, y) \end{pmatrix}. \tag{2.2.5}$$

Here $\boldsymbol{x}^0 = (x_1, \ldots, x_n) \in X^n$ and $\boldsymbol{y}^0 = (y_1, \ldots, y_n) \in Y^n$ are the n-tuples from (2.2.4) *and $C_n^{-1} h$ stands for the inverse of the Casorati matrix $C_n h$.*

Proof: (i) If h is of the form (2a) on $X \times Y$, then each of the $n+1$ functions

$$h(-, y_1), h(-, y_2), \ldots, h(-, y_{n+1}) \tag{2.2.6}$$

is a linear combination of the components $f_1, f_2, \ldots, f_n$ in the set X, for each fixed elements $y_1, \ldots, y_{n+1} \in Y^{n+1}$. (Notice that (2.2.6) is an example of (1.1.2) with functionals $L_i g = g(y_i)$.) Thus, the $(n+1)$ functions (2.2.6) are linearly dependent in X and Theorem 1.4.1 implies that the Casorati determinant of the $(n+1)$-tuple (2.2.6) (equal to the $(n+1)$-th order determinant $\det C_{n+1} h(-, \boldsymbol{y})$) vanishes at each point of X^{n+1}. This proves the conclusion (2.2.3).

(ii) Assume that h satisfies (2.2.3) and (2.2.4) for some integer $n \geq 1$. Let us fix the n-tuples $\boldsymbol{x}^0 = (x_1, x_2, \ldots, x_n)$ and $\boldsymbol{y}^0 = (y_1, y_2, \ldots, y_n)$ from (2.2.4) and utilize the equality from (2.2.3) with $(n+1)$-tuples

$$\boldsymbol{x} = (x_1, x_2, \ldots, x_n, x) \quad \text{and} \quad \boldsymbol{y} = (y_1, y_2, \ldots, y_n, x)$$

where $x \in X$ and $y \in Y$ are arbitrary. We get the equality

$$\begin{vmatrix} h(x_1,y_1) & h(x_1,y_2) & \dots & h(x_1,y_n) & h(x_1,y) \\ h(x_2,y_1) & h(x_2,y_2) & \dots & h(x_2,y_n) & h(x_2,y) \\ \vdots & \vdots & \ddots & \vdots & \vdots \\ h(x_n,y_1) & h(x_n,y_2) & \dots & h(x_n,y_n) & h(x_n,y) \\ h(x,y_1) & h(x,y_2) & \dots & h(x,y_n) & h(x,y) \end{vmatrix} = 0\,.$$

Computing the determinant with respect to the last row, we obtain

$$\sum_{i=1}^{n}(-1)^{n+i}h(x,y_i)\varphi_i(y) + h(x,y)\cdot \det C_n h(\boldsymbol{x}^0,\boldsymbol{y}^0) = 0 \quad (x\in X,\ y\in Y) \tag{2.2.7}$$

where the functions $\varphi_i : Y \to \mathbb{K}$ are given by

$$\varphi_i(y) := \begin{vmatrix} h(x_1,y_1) & \dots & h(x_1,y_{i-1}) & h(x_1,y_{i+1}) & \dots & h(x_1,y_n) & h(x_1,y) \\ h(x_2,y_1) & \dots & h(x_2,y_{i-1}) & h(x_2,y_{i+1}) & \dots & h(x_2,y_n) & h(x_2,y) \\ \vdots & \ddots & \vdots & \vdots & \ddots & \vdots & \vdots \\ h(x_n,y_1) & \dots & h(x_n,y_{i-1}) & h(x_n,y_{i+1}) & \dots & h(x_n,y_n) & h(x_n,y) \end{vmatrix}$$

for $i = 1, 2, \dots, n$. Computing now these determinants for φ_i with respect to the last column and taking in account that $\det C_n h(\boldsymbol{x}^0, \boldsymbol{y}^0) \neq 0$, we compute $h(x,y)$ from (2.2.7) in the form

$$h(x,y) = \sum_{i=1}^{n} h(x,y_i)\Big(\sum_{j=1}^{n}\alpha_{ij}h(x_i,y)\Big) \quad (x\in X,\ y\in Y) \tag{2.2.8}$$

where $\alpha_{ij} \in \mathbb{K}$ are n^2 suitable constants. Notice that (2.2.8) is a decomposition of type (2a) that can be rewritten in the matrix-product form

$$h(x,y) = \big(h(x,y_1), h(x,y_2), \dots, h(x,y_n)\big)\cdot A \cdot \begin{pmatrix} h(x_1,y) \\ h(x_2,y) \\ \vdots \\ h(x_n,y) \end{pmatrix} \tag{2.2.9}$$

in which A is the $n \times n$ matrix with entries α_{ij}, $i, j \in \{1, 2, \dots, n\}$. To verify the validity of (2.2.5), we have only to show that $A = C_n^{-1}h(\boldsymbol{x}^0, \boldsymbol{y}^0)$. This can be done very easily: the representation (2.2.9) yields the matrix identity

$$C_n h(\boldsymbol{x},\boldsymbol{y}) = C_n h(\boldsymbol{x},\boldsymbol{y}^0)\cdot A \cdot C_n h(\boldsymbol{x}^0,\boldsymbol{y}) \qquad (\boldsymbol{x}\in X^n,\ \boldsymbol{y}\in Y^n).$$

Setting here $\boldsymbol{x}=\boldsymbol{x}^0$ and $\boldsymbol{y}=\boldsymbol{y}^0$, we find that $A = C_n^{-1}h(\boldsymbol{x}^0,\boldsymbol{y}^0)$. It remains to prove that both n-tuples of components in (2.2.8) are linearly independent. However, this fact immediately follows from Theorem 1.4.1, because the matrix A is nonsingular and $\det C_n^{-1}h(\boldsymbol{x}^0,\boldsymbol{y}^0)$ is the common nonzero value of the Casorati determinants for the n-tuples $\{h(-,y_i)\}_{i=1}^n$ and $\{h(x_i,-)\}_{i=1}^n$. □

Remarks 2.2.2: **(i)** It is easily seen from the above proof that the sufficient conditions for h to have a decomposition (2a) can be stated in a slightly weaker form. In fact, we need (2.2.3) to be valid only for the $(n+1)$-tuples $\boldsymbol{x}$ and $\boldsymbol{y}$ of the special form

$$\boldsymbol{x} = (x_1, x_2, \ldots, x_n, x) \quad \text{and} \quad \boldsymbol{y} = (y_1, y_2, \ldots, y_n, x)$$

where $\boldsymbol{x}^0 = (x_1,\ldots,x_n) \in X^n$ and $\boldsymbol{y}^0 = (y_1,\ldots,y_n) \in Y^n$ are any *fixed* n-tuples satisfying (2.2.4), while the last coordinates x and y assume values freely on the sets X and Y, respectively.

(ii) The known rank-property of square matrices implies that if a function h satisfies (2.2.3) for some $n=n_0$, then it does so for each $n \geq n_0$ as well. Taking the smallest integer n for which (2.2.3) is fulfilled, we ensure that the condition (2.2.4) holds automatically. This is why the additional condition (2.2.4) makes essentially no restriction in applying Theorem 2.2.1. (Compare this with the part (ii) of Remarks 2.1.2 concerning the converse part of Theorem 2.2.1.)

2.3. Minimal decompositions

Let us start this section by the following simple observation. Suppose that a function h has a decomposition

$$h(x,y) = \sum_{i=1}^{n} f_i(x)g_i(y) \qquad (x \in X,\ y \in Y) \tag{2.3.1}$$

in which an n-tuple of functions $\{f_i\}_{i=1}^n$ or $\{g_i\}_{i=1}^n$ is linearly dependent (in the set X or Y, respectively). For instance, assume that

$$f_i(x) = \sum_{j=1}^{m} \alpha_{ij}\tilde{f}_j(x) \qquad (x \in X,\ 1 \leq i \leq n) \tag{2.3.2}$$

where $\alpha_{ij} \in \mathbb{K}$ are constants and $m<n$. Substituting (2.3.2) into the right-hand side of (2.3.1), we get a new decomposition

$$h(x,y) = \sum_{j=1}^{m} \tilde{f}_j(x)\tilde{g}_j(y) \qquad (x \in X,\ y \in Y) \tag{2.3.3}$$

with $\tilde{g}_j = \sum_{i=1}^{m} \alpha_{ij} g_i$, $1 \le j \le m$. Since $m < n$, we consider (2.3.3) to be a *reduction* of (2.3.1). Now, if the functions $\{\tilde{g}_j\}_{j=1}^{m}$ are linearly dependent again, we can repeat this procedure of reduction. After a finite number of such repetitions, we obtain such a decomposition of type (2.3.1) in which both n-tuples $\{f_i\}_{i=1}^{n}$ and $\{g_i\}_{i=1}^{n}$ are linearly independent. (In fact, the last assertion fails to hold if h vanishes at each point of $X \times Y$. However, in this case we can formally consider (2.3.1) with $n=0$ as an *empty* sum.)

The preceding observation leads us to the following: We call (2.3.1) a *minimal decomposition* of the function h (in the set $X \times Y$) if and only if both the n-tuples $\{f_i\}_{i=1}^{n}$ and $\{g_i\}_{i=1}^{n}$ are linearly independent in the sets X and Y, respectively.

Theorem 2.3.1 [N 1]: *Let*

$$h(x,y) = \sum_{i=1}^{n} f_i(x) g_i(y) \quad (x \in X,\ y \in Y) \tag{2.3.4a}$$

$$h(x,y) = \sum_{i=1}^{\tilde{n}} \tilde{f}_i(x) \tilde{g}_i(y) \quad (x \in X,\ y \in Y) \tag{2.3.4b}$$

be two arbitrary decompositions of the same function $h : X \times Y \to \mathbb{K}$ and let (2.3.4a) *be a minimal one. Then $\tilde{n} \ge n$ and* (2.3.4b) *is minimal if and only if the equality $\tilde{n} = n$ holds. If this is the case, then there exists a constant nonsingular matrix C of size $n \times n$ such that the equalities*

$$\begin{pmatrix} \tilde{f}_1(x) \\ \tilde{f}_2(x) \\ \vdots \\ \tilde{f}_n(x) \end{pmatrix} = C^T \cdot \begin{pmatrix} f_1(x) \\ f_2(x) \\ \vdots \\ f_n(x) \end{pmatrix} \quad \text{and} \quad \begin{pmatrix} \tilde{g}_1(y) \\ \tilde{g}_2(y) \\ \vdots \\ \tilde{g}_n(y) \end{pmatrix} = C^{-1} \cdot \begin{pmatrix} g_1(y) \\ g_2(y) \\ \vdots \\ g_n(y) \end{pmatrix} \tag{2.3.5}$$

hold for each $x \in X$ and $y \in Y$, respectively. (The matrices C^T and C^{-1} in (2.3.5) *are the transpose and the inverse of C respectively.)*

Proof: If h has a decomposition (2a) on $X \times Y$, then

$$\begin{aligned} C_n h(\boldsymbol{x}, \boldsymbol{y}) &= \left[\sum_{k=1}^{n} f_k(x_i) g_k(y_j) \right]_{i,j=1}^{n} = \left[f_k(x_i) \right]_{i,k=1}^{n} \cdot \left[g_k(y_j) \right]_{k,j=1}^{n} \\ &= C(f_1, f_2, \ldots, f_n; \boldsymbol{x}) \cdot C^T(g_1, g_2, \ldots, g_n; \boldsymbol{y}) \end{aligned} \tag{2.3.6}$$

for each $\boldsymbol{x} = (x_1, \ldots, x_n) \in X^n$ and each $\boldsymbol{y} = (y_1, \ldots, y_n) \in Y^n$. Thus the identity

$$\det C_n h(\boldsymbol{x}, \boldsymbol{y}) = \det C(f_1, f_2, \ldots, f_n; \boldsymbol{x}) \cdot \det C(g_1, g_2, \ldots, g_n; \boldsymbol{y}) \tag{2.3.7}$$

holds on $X^n \times Y^n$. By Theorem 1.4.1, the n-tuples of functions $\{f_i\}_{i=1}^n$ and $\{g_i\}_{i=1}^n$ are linearly independent (in X and Y, respectively) if and only if the Casorati determinants from the right-hand side of (2.3.7) have at least one nonzero value in X^n and Y^n, respectively. In view of (2.3.7), the last condition is exactly (2.2.4). Consequently, Theorem 2.2.1 leads to the conclusion that a given decomposition (2a) is minimal if and only if the number n of the terms in (2a) is equal to the (unique) number n satisfying (2.2.3) and (2.2.4) simultaneously. Moreover, if (2.3.4b) is any (not necessarily minimal) decomposition, then the first part of Theorem 2.2.1 implies that $\det C_{\tilde{n}+1}h(\boldsymbol{x},\boldsymbol{y}) = 0$ for each $\boldsymbol{x} \in X^{\tilde{n}+1}$ and $\boldsymbol{y} \in Y^{\tilde{n}+1}$, and hence $\tilde{n} \geq n$, where n is the number of terms in the minimal decomposition (2.3.4a). Supposing now $\tilde{n} = n$ and choosing the n-tuples $\boldsymbol{x}^0 = (x_1, \ldots, x_n) \in X^n$ and $\boldsymbol{y}^0 = (y_1, \ldots, y_n) \in Y^n$ from (2.2.4), we obtain from (2.3.4) the following vector identity

$$C(g_1,\ldots,g_n;\boldsymbol{y}^0)\cdot\begin{pmatrix} f_1(x)\\ f_2(x)\\ \vdots\\ f_n(x)\end{pmatrix} = C(\tilde{g}_1,\ldots,\tilde{g}_n;\boldsymbol{y}^0)\cdot\begin{pmatrix} \tilde{f}_1(x)\\ \tilde{f}_2(x)\\ \vdots\\ \tilde{f}_n(x)\end{pmatrix} \qquad (x\in X). \tag{2.3.8}$$

Notice also that (2.3.6) yields the equality

$$C(f_1,\ldots,f_n;\boldsymbol{x}^0)\cdot C^T(g_1,\ldots,g_n;\boldsymbol{y}^0) = C(\tilde{f}_1,\ldots,\tilde{f}_n;\boldsymbol{x}^0)\cdot C^T(\tilde{g}_1,\ldots,\tilde{g}_n;\boldsymbol{y}^0)\,. \tag{2.3.9}$$

In view of (2.2.4) and (2.3.7), all four matrices in (2.3.9) are nonsingular. Hence the equality (2.3.8) implies the first part of (2.3.5) with a matrix C satisfying

$$C = C^T(g_1,\ldots,g_n;\boldsymbol{y}^0)\cdot C^{T^{-1}}(\tilde{g}_1,\ldots,\tilde{g}_n;\boldsymbol{y}^0)\,. \tag{2.3.10}$$

Analogously, the vector equality

$$C(f_1,\ldots,f_n;\boldsymbol{x}^0)\cdot\begin{pmatrix} g_1(y)\\ g_2(y)\\ \vdots\\ g_n(y)\end{pmatrix} = C(\tilde{f}_1,\ldots,\tilde{f}_n;\boldsymbol{x}^0)\cdot\begin{pmatrix} \tilde{g}_1(y)\\ \tilde{g}_2(y)\\ \vdots\\ \tilde{g}_n(y)\end{pmatrix} \qquad (y\in Y)$$

forces the validity of the second part of (2.3.5), with a matrix

$$C = C^{-1}(f_1,\ldots,f_n;\boldsymbol{x}^0)\cdot C(\tilde{f}_1,\ldots,\tilde{f}_n;\boldsymbol{x}^0)\,. \tag{2.3.11}$$

It follows from (2.3.9) that (2.3.10) and (2.3.11) determine the same matrix C. This completes the proof. □

Corollary 2.3.2: (i) *Under the conditions* (2.1.2) *and* (2.1.3), *each minimal decomposition of the function* h *is of the form*

$$h(x,y) = \sum_{i=1}^{n} \Big(\sum_{j=1}^{n} \alpha_{ij} h_{y^{j-1}}(x, y_0) \Big) \cdot \Big(\sum_{j=1}^{n} \beta_{ij} h_{x^{j-1}}(x_0, y) \Big) \tag{2.3.12}$$

where $A = [\alpha_{ij}]_{i,j=1}^{n}$ *and* $B = [\beta_{ij}]_{i,j=1}^{n}$ *are any constant nonsingular matrices satisfying* $A^T \cdot B = W_n^{-1} h(x_0, y_0)$.

(ii) *Under the conditions* (2.2.3) *and* (2.2.4), *each minimal decomposition of the function* h *is of the form*

$$h(x,y) = \sum_{i=1}^{n} \Big(\sum_{j=1}^{n} \alpha_{ij} h(x, y_j) \Big) \cdot \Big(\sum_{j=1}^{n} \beta_{ij} h(x_j, y) \Big) \tag{2.3.13}$$

where $\boldsymbol{x}^0 = (x_1, \ldots, x_n) \in X^n$ *and* $\boldsymbol{y}^0 = (y_1, \ldots, y_n) \in Y^n$ *are the* n*-tuples from* (2.2.4) *and* $A = [\alpha_{ij}]_{i,j=1}^{n}$ *and* $B = [\beta_{ij}]_{i,j=1}^{n}$ *are any constant nonsingular matrices satisfying* $A^T \cdot B = C_n^{-1} h(\boldsymbol{x}^0, \boldsymbol{y}^0)$.

Proof: In view of Theorem 2.3.1, the formulas (2.3.12) and (2.3.13) immediately follow from (2.1.11) and (2.2.5), respectively. □

Remark 2.3.3: The representation formula (2.3.13) has an important consequence: If a function h of type (2a) possesses some *regularity property* (like continuity, periodicity, smootheness of some order, L^p-integrability, etc.) in one of the variables x, y (say x), then the same property is possessed by each component f_i of any minimal decomposition (2a) of the function h.

Notice that the statement of Theorem 2.2.1, the general criterion of the decomposability (2a), involves the Casorati determinants of two consecutive orders, namely $\det C_n h$ and $\det C_{n+1} h$. The following result shows that the same criterion can be stated without using the determinant $\det C_{n+1} h$, in the form of the following functional equation

$$\det C_n h(\boldsymbol{x}, \boldsymbol{y}) \cdot \det C_n h(\boldsymbol{u}, \boldsymbol{v}) = \det C_n h(\boldsymbol{x}, \boldsymbol{v}) \cdot \det C_n h(\boldsymbol{u}, \boldsymbol{y}) \tag{2.3.14}$$

for the Casorati determinant $\det C_n h$.

Theorem 2.3.4 [Ši 1]: *Let* $h\colon X \times Y \to \mathbb{K}$ *be a function, where* X *and* Y *are arbitrary nonempty sets. If* h *is of the form* (2a) *on the set* $X \times Y$, *then the equality* (2.3.14) *holds for each* n*-tuples* $\boldsymbol{x}, \boldsymbol{u} \in X^n$ *and* $\boldsymbol{y}, \boldsymbol{v} \in Y^n$. *Conversely, if* (2.3.14) *holds for each* $\boldsymbol{x} \in X^n$, $\boldsymbol{y} \in Y^n$ *and for some* $\boldsymbol{u} \in X^n$, $\boldsymbol{v} \in Y^n$ *satisfying* $\det C_n h(\boldsymbol{u}, \boldsymbol{v}) \neq 0$, *then* h *is of the form* (2a) *on the set* $X \times Y$.

Proof: If h is of the form (2a) on the set $X \times Y$, then the formula (2.3.7) implies that both sides of (2.3.14) are equal to the same product of the four determinants of Casorati matrices

$$C(f_1, \dots, f_n; \boldsymbol{x}),\ C(g_1, \dots, g_n; \boldsymbol{y}),\ C(f_1, \dots, f_n; \boldsymbol{u}) \quad \text{and} \quad C(g_1, \dots, g_n; \boldsymbol{v}).$$

Conversely, suppose that $\det C_n h(\boldsymbol{u}, \boldsymbol{v}) \neq 0$ for some (fixed) $\boldsymbol{u} \in Y^n$ and $\boldsymbol{v} \in Y^n$ and that (2.3.14) holds for each $\boldsymbol{x} \in X^n$ and $\boldsymbol{y} \in Y^n$. If $\boldsymbol{v} = (v_1, \dots, v_n)$, then Theorem 1.4.1 implies that the n-tuple of functions

$$\mathscr{S}_1 = \{h(-, v_1), h(-, v_2), \dots, h(-, v_n)\}$$

is linearly independent in the set X, because the Casorati determinant of $\mathscr{S}_1$ (equal to $\det C_n h(\boldsymbol{x}, \boldsymbol{v})$) is nonzero at the point $\boldsymbol{x} = \boldsymbol{u}$. Let us verify that for each (fixed) $y \in Y$, the function $h(-, y)$ is a linear combination of the n-tuple $\mathscr{S}_1$ in the set X. We need to show that

$$h(x, y) = \sum_{i=1}^{n} g_i(y) h(x, v_i) \qquad (x \in X) \tag{2.3.15}$$

with suitable "coefficients" $g_i \colon Y \to \mathbb{K}$. Suppose, on the contrary, that the system

$$\mathscr{S}_2 = \{h(-, y), h(-, v_2), \dots, h(-, v_n)\}$$

is linearly independent in X, for some $y \in Y$ (otherwise (2.3.15) clearly holds with $g_1(y) = 0$). Let us fix such a $y \in Y$ and put $\boldsymbol{y} = (y, v_2, v_3, \dots, v_n)$ in (2.3.14). Since $\det C_n h(\boldsymbol{x}, \boldsymbol{y})$ is then the Casorati determinant of $\mathscr{S}_2$, and since

$$\det C_n h(\boldsymbol{x}, \boldsymbol{y}) = \beta \cdot \det C_n h(\boldsymbol{x}, \boldsymbol{v}) \qquad \left(\beta = \frac{\det C_n h(\boldsymbol{u}, \boldsymbol{y})}{\det C_n h(\boldsymbol{u}, \boldsymbol{v})},\ \boldsymbol{x} \in X^n\right)$$

where $\beta \neq 0$ is independent of $\boldsymbol{x} \in X^n$, then Corollary 1.4.5 implies that $\mathscr{S}_1$ and $\mathscr{S}_2$ are two bases of the same linear space of functions defined on X. Consequently, the function $h(-, y) \in \mathscr{S}_2$ is a linear combination of the n-tuple $\mathscr{S}_1$ (with coefficients depending on y) and the validity of (2.3.15) is established. □

The next result deals with functions in two variables, defined on a Cartesian square X^2 and decomposable into the form

$$h(x, y) = \sum_{i,j=1}^{n} \alpha_{ij} f_i(x) f_j(y) \qquad (x, y \in X) \tag{2.3.16}$$

with a suitable n-tuple of functions $f_i\colon X \to \mathbb{K}$ and n^2 constants $\alpha_{ij} \in \mathbb{K}$. Without loss of generality, we may assume that the n-tuple of components $\{f_i\}_{i=1}^n$ is linearly independent in the set X and that the constant square matrix $A := [\alpha_{ij}]_{i,j=1}^n$ is nonsingular, otherwise the number n in (2.3.16) can be reduced. This assumption is equivalent to the fact that the n-th order Casorati determinant $\det C_n h$ satisfies

$$\det C_n h(\boldsymbol{z}, \boldsymbol{z}) \neq 0 \quad \text{for some } \boldsymbol{z} \in X^n \tag{2.3.17}$$

because the representation (2.3.16) leads to the equality

$$\det C_n h(\boldsymbol{x}, \boldsymbol{y}) = \det C(f_1, \ldots, f_n; \boldsymbol{x}) \cdot \det A \cdot \det C(f_1, \ldots, f_n; \boldsymbol{y}) \quad (\boldsymbol{x}, \boldsymbol{y} \in X^n) \tag{2.3.18}$$

and hence the condition (2.3.17) follows from Theorem 1.4.1.

Theorem 2.3.5 [Ši 1]: *Let $h\colon X^2 \to \mathbb{K}$ be a function, where X is a nonempty set. If h has a decomposition* (2.3.16), *then the equality*

$$\det C_n h(\boldsymbol{x}, \boldsymbol{y}) \cdot \det C_n h(\boldsymbol{z}, \boldsymbol{z}) = \det C_n h(\boldsymbol{x}, \boldsymbol{z}) \cdot \det C_n h(\boldsymbol{y}, \boldsymbol{z}) \tag{2.3.19}$$

holds for each $\boldsymbol{x}, \boldsymbol{y}, \boldsymbol{z} \in X^n$. Conversely, if h satisfies (2.3.17) *and the equality* (2.3.19) *holds for each $\boldsymbol{x}, \boldsymbol{y} \in X^n$ and for a fixed $\boldsymbol{z} = (z_1, z_2, \ldots, z_n) \in X^n$ taken from* (2.3.17), *then h is of the form* (2.3.16). *Moreover, the components f_i in* (2.3.16) *can be then chosen in such a manner for the constants α_{ij} to be equal to*

$$a_{ij} = h(z_i, z_j) \qquad (i, j \in \{1, 2, \ldots, n\}) \tag{2.3.20}$$

where $(z_1, z_2, \ldots, z_n)$ is the n-tuple $\boldsymbol{z}$ from (2.3.17). *Especially, the function h is then symmetric in the variables x, y if and only if the equality $h(z_i, z_j) = h(z_j, z_i)$ holds for any $i, j \in \{1, 2, \ldots, n\}$.*

Proof: If $h\colon X^2 \to \mathbb{K}$ is of the form (2.3.16), then the equality (2.3.19) immediately follows from (2.3.18). Conversely, suppose that (2.3.19) holds for each pair $\boldsymbol{x}, \boldsymbol{y} \in X^n$ and for some (fixed) $\boldsymbol{z} = (z_1, z_2, \ldots, z_n) \in X^n$ from (2.3.17). Then

$$\begin{aligned} \det C_n h(\boldsymbol{x}, \boldsymbol{y}) &= \frac{\det C_n h(\boldsymbol{x}, \boldsymbol{z}) \cdot \det C_n h(\boldsymbol{y}, \boldsymbol{z})}{\det C_n h(\boldsymbol{z}, \boldsymbol{z})} \\ &= \frac{\det C_n h(\boldsymbol{y}, \boldsymbol{z}) \cdot \det C_n h(\boldsymbol{x}, \boldsymbol{z})}{\det C_n h(\boldsymbol{z}, \boldsymbol{z})} = \det C_n h(\boldsymbol{y}, \boldsymbol{x}) \end{aligned}$$

which means that $\det C_n h(\boldsymbol{x}, \boldsymbol{y})$ is symmetric in the variables $\boldsymbol{x}, \boldsymbol{y}$. Due to this fact, (2.3.19) can be rewritten into the form

$$\det C_n h(\boldsymbol{x}, \boldsymbol{y}) \cdot \det C_n h(\boldsymbol{z}, \boldsymbol{z}) = \det C_n h(\boldsymbol{x}, \boldsymbol{z}) \cdot \det C_n h(\boldsymbol{z}, \boldsymbol{y})$$

which is condition (2.3.14) with $\boldsymbol{u} = \boldsymbol{v} = \boldsymbol{z}$. According to Theorem 2.3.4, there exist two n-tuples of functions $p_i: X \to \mathbb{K}$ and $q_i : X \to \mathbb{K}$ $(i = 1, \ldots, n)$ such that our function h is of the form

$$h(x, y) = \sum_{i=1}^{n} p_i(x) q_i(y) \qquad (x, y \in X). \tag{2.3.21}$$

Consequently, the Casorati determinant $\det C_n h$ possesses the property (2.3.7):

$$\det C_n h(\boldsymbol{x}, \boldsymbol{y}) = P(\boldsymbol{x}) \cdot Q(\boldsymbol{y}),$$

where P and Q are the Casorati determinants of the n-tuples $\{p_i\}_{i=1}^{n}$ and $\{q_i\}_{i=1}^{n}$, respectively. Notice that the determinants P and Q satisfy

$$P(\boldsymbol{x}) Q(\boldsymbol{y}) = P(\boldsymbol{y}) Q(\boldsymbol{x}) \quad \text{and} \quad P(\boldsymbol{z}) Q(\boldsymbol{z}) \neq 0$$

because of the mentioned symmetry of $\det C_n h$ and (2.3.17). This leads to the conclusion that

$$P(\boldsymbol{x}) = \beta \cdot Q(\boldsymbol{x}) \qquad \left(\beta = \frac{P(\boldsymbol{z})}{Q(\boldsymbol{z})} \neq 0, \ \boldsymbol{x} \in X^n \right).$$

In view of Corollary 1.4.5, both n-tuples $\{p_i\}_{i=1}^{n}$ and $\{q_i\}_{i=1}^{n}$ are two bases of the same linear space of functions on X. Since $P(\boldsymbol{z}) \neq 0$, Theorem 1.4.2 ensures that this linear space has a basis $\{f_i\}_{i=1}^{n}$ satisfying n^2 conditions

$$f_i(z_j) = \delta_{ij} \qquad (i, j \in \{1, 2, \ldots, n\}), \tag{2.3.22}$$

where δ is the Kronecker symbol. Substituting p_i and q_i in (2.3.21) by the corresponding linear combinations of f_i, we obtain a decomposition (2.3.16), in which the coefficients a_{ij} are as in (2.3.20), because of (2.3.22). □

2.4. Matrix factorizations

We will finish this chapter by finding functional and differential equations for matrix-valued functions $H = H(x, y)$ admitting the factorization

$$H(x, y) = F(x) \cdot G(y) \qquad (x \in X, \ y \in Y) \tag{2.4.1}$$

where $\cdot$ denotes the usual matrix multiplication. Suppose that matrices H, F and G in (2.4.1) are of size $n \times n$ and denote their entries by h_{ij}, f_{ij} and g_{ij}, respectively, where $i, j \in \{1, 2, \ldots, n\}$. Then (2.4.1) represents a system of n^2 scalar equalities

$$h_{ij}(x,y) = \sum_{k=1}^{n} f_{ik}(x) g_{kj}(y) \,,$$

each of them being of type (2a). Consequently, the fundamental Theorem 2.1.1 yields a necessary (but not sufficient) condition for a (smooth) matrix function H to have factorization (2.4.1): *each entry h_{ij} is a solution of the Wronski equation* (2.1.2). We will show here that criteria for factorization (2.4.1) can be stated in terms of matrix operations, without taking single entries of the matrix H and without using equations like (2.1.2). However, our procedure (described in [Ši 4]) is not applicable unless the values of H are nonsingular, i.e. $\det H(x,y) \neq 0$ for all x and y. As usual, we will denote by $GL_n(\mathbb{K})$ the group of all nonsingular matrices of size $n \times n$ with elements from the field $\mathbb{K}$. Let us finish this introductory part by remarking that a smooth matrix H of type (2.4.1) need not satisfy the equation

$$H_{xy} \cdot H - H_x \cdot H_y = 0 \,, \tag{2.4.2}$$

a formal matrix analogue of (2.1.2) with $n = 1$. (Equation (2.4.2) holds if the matrices F and G in (2.4.1) commute, which is rather an *exceptional* case.) The correct version of (2.4.2) is given in Theorem 2.4.5 below (see (2.4.8)).

First we derive a functional equation that characterizes functions (2.4.1) without any smoothness condition.

Theorem 2.4.1: *Let $H : X \times Y \to GL_n(\mathbb{K})$ be a mapping, where X and Y are arbitrary nonempty sets. Choose elements $x_1 \in X$ and $y_1 \in Y$. Then the mapping H has a factorization* (2.4.1) *if and only if it satisfies the functional equation*

$$H(x,y) = H(x,y_1) \cdot H^{-1}(x_1,y_1) \cdot H(x_1,y) \quad \text{for each } x \in X \text{ and } y \in Y \,. \tag{2.4.3}$$

Moreover, the factors $F : X \to GL_n(\mathbb{K})$ and $G : Y \to GL_n(\mathbb{K})$ from any representation (2.4.1) *are exactly pairs of the form*

$$F(x) = H(x,y_1) \cdot C \quad \text{and} \quad G(y) = D \cdot H(x_1,y) \tag{2.4.4}$$

where $C, D \in GL_n(\mathbb{K})$ are arbitrary constant matrices satisfying the equality $C \cdot D = H^{-1}(x_1,y_1)$.

Proof: Let H be as in (2.4.1). Setting first $y = y_1$ and then $x = x_1$ in (2.4.1), we find that

$$F(x) = H(x,y) \cdot G^{-1}(y_1) \text{ and } G(y) = F^{-1}(x_1) \cdot H(x_1,y)$$

for each $x \in X$ and $y \in Y$. Multiplying these equalities and taking into account that

$$G^{-1}(y_1) \cdot F^{-1}(x_1) = (F(x_1) \cdot G(y_1))^{-1} = H^{-1}(x_1,y_1) \,,$$

we conclude that H satisfies (2.4.3) and that (2.4.4) holds. Conversely, if H satisfies (2.4.3) and if $C, D \in GL_n(\mathbb{K})$ are arbitrary matrices satisfying $C \cdot D = H^{-1}(x_1, y_1)$, then

$$\big(H(x, y_1)\, C\big) \cdot \big(D\, H(x_1, y)\big) = H(x, y_1) \cdot H^{-1}(x_1, y_1) \cdot H(x_1, y) = H(x, y)\,,$$

and the proof is complete. □

Remark 2.4.2: Let us add to Theorem 2.4.1 a simple but important rule:

$$H \textit{ is of type } (2.4.1) \;\Rightarrow\; H(x_1, y) \cdot H^{-1}(x_2, y) \textit{ does not depend on } y\,, \qquad (2.4.5)$$

which will be used in Chapter 3.

Now we turn our attention to the matrix functions H of type (2.4.1) which are differentiable in one or both variables, say x in Theorem 2.4.3 (for the case of the variable y, see Remark 2.4.4). We will show that such functions are characterized by an equation which is differential in x and functional in y.

Theorem 2.4.3: *Let $H : X \times Y \to GL_n(\mathbb{K})$ be a mapping, where X is an interval in $\mathbb{R}$ and Y is a nonempty set. Suppose that the partial derivative H_x exists at each point of $X \times Y$. Then the mapping H has a factorization* (2.4.1) *if and only if it satisfies*

$$H_x(x, y) \cdot H^{-1}(x, y) = H_x(x, y_1) \cdot H^{-1}(x, y_1) \quad \textit{for each } x \in X \textit{ and } y, y_1 \in Y. \qquad (2.4.6)$$

Proof: (i) If H is as in (2.4.1), then

$$H_x(x, y) \cdot H^{-1}(x, y) = \big(F'(x)G(y)\big) \cdot \big(G^{-1}(y)F^{-1}(x)\big) = F'(x) \cdot F^{-1}(x)$$

for each $y \in Y$; hence both sides of (2.4.6) are equal to $F'(x) \cdot F^{-1}(x)$.

(ii) If H satisfies (2.4.6), then

$$\begin{aligned}
&\frac{\partial}{\partial x}\big(H^{-1}(x, y_1)H(x, y)\big)\\
&= -H^{-1}(x, y_1)H_x(x, y_1)H^{-1}(x, y_1)H(x, y) + H^{-1}(x, y_1)H_x(x, y)\\
&= H^{-1}(x, y_1)\big[-H_x(x, y_1)H^{-1}(x, y_1) + H_x(x, y)H^{-1}(x, y)\big]H(x, y) = 0\,.
\end{aligned}$$

Thus $H^{-1}(x, y_1) \cdot H(x, y)$ does not depend on $x \in X$, i.e.

$$H^{-1}(x, y_1) \cdot H(x, y) = H^{-1}(x_1, y_1) \cdot H(x_1, y) \text{ for each } x \in X$$

where $x_1 \in X$ is a chosen point. Multiplying the last equality by $H(x, y_1)$ from the left, we obtain the factorization (2.4.3). □

Remark 2.4.4: The reader can easily verify that

$$H^{-1}(x,y) \cdot H_y(x,y) = H^{-1}(x_1,y) \cdot H_y(x_1,y) \quad (x, x_1 \in X,\ y \in Y) \tag{2.4.7}$$

by analogy to (2.4.6) for functions H differentiable in the variable y.

Now we state a differential criterion of (2.4.1) for mappings H which are smooth in both variables x and y.

Theorem 2.4.5: *Let $H : X \times Y \to GL_n(\mathbb{K})$ be a mapping, where X and Y are two intervals in $\mathbb{R}$. Suppose that the partial derivatives H_x, H_y and $H_{xy} = (H_x)_y$ exist at each point of $X \times Y$. Then the mapping H has a factorization* (2.4.1) *if and only if it solves the differential equation*

$$H_{xy} = H_x \cdot H^{-1} \cdot H_y \quad \text{on the rectangle } X \times Y. \tag{2.4.8}$$

Proof: If H is as in (2.4.1) and the derivatives H_x and H_y exist, then (2.4.4) implies that the derivatives $F' = \frac{dF}{dx}$ and $G' = \frac{dG}{dy}$ also exist. So we can write

$$\begin{aligned} H_x \cdot H^{-1} \cdot H_y &= (F'G) \cdot (FG)^{-1} \cdot (FG') = F'GG^{-1}F^{-1}FG' \\ &= F'G' = H_{xy} \end{aligned}$$

which means that H satisfies (2.4.8). Conversely, let H be such that the derivatives H_x, H_y, $H_{xy} = (H_x)_y$ exist and satisfy (2.4.8). Then the product $H_x \cdot H^{-1}$ is differentiable in y and

$$\begin{aligned} \frac{\partial}{\partial y}\left(H_x \cdot H^{-1}\right) &= H_{xy}H^{-1} - H_xH^{-1}H_yH^{-1} \\ &= (H_{xy} - H_xH^{-1}H_y)H^{-1} = 0 \end{aligned}$$

on the set $X \times Y$. Hence $H_x \cdot H^{-1}$ does not depend on $y \in Y$, i.e. the mapping H satisfies (2.4.6). In view of Theorem 2.4.3, H has a factorization (2.4.1). □

Remark 2.4.6: In the statement of Theorem 2.4.5, the mixed derivative $(H_x)_y$ can be replaced by $(H_y)_x$, because any solution of $(H_y)_x = H_x \cdot H^{-1} \cdot H_y$ satisfies (2.4.7).

As an application of Theorems 2.4.1 and 2.4.5, we now derive functional and differential equations for matrix functions of the form

$$H(x,y) = F(x+y) \cdot G(x-y)\,. \tag{2.4.9}$$

Corollary 2.4.7: (i) *Let S be an abelian group divisible by 2. A given mapping $H : S \times S \to GL_n(\mathbb{K})$ has a factorization* (2.4.9) *if and only if it satisfies the equation*

$$H(x, y) = H\left(\frac{x+y}{2}, \frac{x+y}{2}\right) \cdot H^{-1}(0,0) \cdot H\left(\frac{x-y}{2}, \frac{y-x}{2}\right)$$

for each $x, y \in S$.

(ii) *Suppose that a mapping $H : \mathbb{R} \times \mathbb{R} \to GL_n(\mathbb{K})$ has the second order differential $d^2 H$ at each point of the plane $\mathbb{R} \times \mathbb{R}$. Then H has a factorization* (2.4.9) *if and only if H solves the differential equation*

$$H_{xx} - H_{yy} = (H_x + H_y) \cdot H^{-1} \cdot (H_x - H_y) \quad \text{on } \mathbb{R} \times \mathbb{R}\,.$$

Proof: Parts (i) and (ii) of Corollary 2.4.7 are immediate consequences of Theorems 2.4.1. and 2.4.5 respectively, applied to the following mapping

$$\tilde{H}(u, v) := H\left(\frac{u+v}{2}, \frac{u-v}{2}\right).$$

3 DECOMPOSITIONS OF FUNCTIONS OF SEVERAL VARIABLES

In 1989 Th. M. Rassias posed the question of finding necessary and sufficient conditions for a function h of three variables (say x, y, z) to be represented in the form

$$h(x, y, z) = \sum_{i=1}^{N} e_i(x) f_i(y) g_i(z) \tag{3a}$$

with suitable scalar functions e_i, f_i, g_i (see [Ra 2]). Soon afterwards, this problem was independently posed by Gauchman and Rubel [GR]. One might think that the solution of this problem is a straightforward application of the technique used to solve the problem for functions h of two variables (see Chapter 2). However, this is not the case. A difficulty is that one cannot define determinants of three-dimensional matrices (for example, the Wronski and Casorati determinants). By a three-dimensional matrix of size $m \times n \times p$ we mean an ordered system of elements $[\lambda_{ijk}]_{i=1}^{m}{}_{j=1}^{n}{}_{k=1}^{p}$.

In the following, representations of type (3a) are replaced by more convenient ones

$$h(x, y, z) = \sum_{i=1}^{m} \sum_{j=1}^{n} \sum_{k=1}^{p} \alpha_{ijk} e_i(x) f_j(y) g_k(z) \tag{3b}$$

in which the m-tuple $\{e_i\}$, the n-tuple $\{f_j\}$ and the p-tuple $\{g_k\}$ may be clearly assumed to be linearly independent in their domains of definition. Note first that (3a) can be brought to (3b) by substituting e_i, f_i, g_i in (3a) by linear combinations of elements of some bases of the vector spaces, generated by $\{e_i\}_{i=1}^{N}$, $\{f_i\}_{i=1}^{N}$ and $\{g_i\}_{i=1}^{N}$, respectively. Conversely, each representation (3b) can be obviously rewritten as (3a); as for example with $N = mnp$ (the question of the smallest possible N in the "diagonal" representation (3a) of a given function h is treated in Remark 3.2.6).

We will also solve the generalized problem of decompositions

$$h(x_1,\ldots,x_k)=\sum_{i_1=1}^{m_1}\cdots\sum_{i_k=1}^{m_k}\alpha_{i_1\ldots i_k}f^1_{i_1}(x_1)\cdot\ldots\cdot f^k_{i_k}(x_k) \tag{3c}$$

which was essentially posed by Th. M. Rassias in [Ra 2], and also the problem of the factorization of matrix-valued mappings H into the form

$$H(x_1,x_2,\ldots,x_k)=F_1(x_1)\cdot F_2(x_2)\cdot\ldots\cdot F_k(x_k)\,. \tag{3d}$$

3.1. Method of reduction

In this section we will show that the decomposition problem (3c) for a function h of $k\geq 3$ variables is equivalent to a system of k decomposition problems (2a).

Notice that for each $j=1,2,\ldots,k$, the terms in the right-hand side of (3c) can be rearranged into the form

$$h(x_1,\ldots,x_k)=\sum_{i=1}^{m_j}f^j_i(x_j)\varphi^j_i(x_1,\ldots,x_{j-1},x_{j+1},\ldots,x_k) \tag{3.1.1}$$

where φ^j_i is a function equal to the sum of all terms

$$\alpha_{i_1\ldots i_{j-1}\,i\,i_{j+1}\ldots i_k}f^1_{i_1}(x_1)\cdot\ldots\cdot f^{j-1}_{i_{j-1}}(x_{j-1})\cdot f^{j+1}_{i_{j+1}}(x_{j+1})\cdot\ldots\cdot f^k_{i_k}(x_k) \tag{3.1.2}$$

over all the values of indices $i_1,\ldots,i_{j-1},i_{j+1},\ldots,i_k$ indicated in (3c) . Before we state the converse assertion that the existence of k decompositions (3.1.1) ensures a representation (3c), let us emphasize the fact that each decomposition (3.1.1) is of type (2a), with a pair of independent variables

$$x=x_j \quad\text{and}\quad y=(x_1,\ldots,x_{j-1},x_{j+1},\ldots,x_k)\,. \tag{3.1.3}$$

Theorem 3.1.1 [ČŠ 1]: *Let $h\colon X_1\times X_2\times\ldots\times X_k\to\mathbb{K}$ be a function where X_1, X_2, ..., X_k are arbitrary nonempty sets. Denote by*

$$Y_j=X_1\times\ldots\times X_{j-1}\times X_{j+1}\times\ldots\times X_k \quad (j=1,2,\ldots,k)$$

and suppose that for each $j=1,\ldots,k$ the function h is of the form (3.1.1) *on the set $X_j\times Y_j$, for some $m_j\geq 1$. If the m_j-tuple of functions $\{f^j_i\}_{i=1}^{m_j}$ in* (3.1.1) *is linearly independent in the set X_j for each $j=1,\ldots,k$, then there exist constants $\alpha_{i_1i_2\ldots i_k}\in\mathbb{K}$ such that the equality* (3c) *holds on the set $X_1\times X_2\times\ldots\times X_k$.*

The proof of Theorem 3.1.1 will rely on the following:

Lemma 3.1.2: (i) *Let* $h : X \times Y \to \mathbb{K}$ *be of the form*

$$h(x,y) = \sum_{i=1}^{n} f_i(x)g_i(y) \quad (x \in X,\ y \in Y) \tag{3.1.4}$$

in which the n*-tuple of functions* $\{f_i\}$ *is linearly independent in the set* X*. Then there exist elements* $u_1, \ldots, u_n \in X$ *such that each function* g_i *from* (3.1.4) *is a linear combination of* n *functions* $h(u_1, -), h(u_2, -), \ldots, h(u_n, -)$ *in the set* Y.

(ii) *Let the* m_j*-tuple of functions* $\{f_i^j\}_{i=1}^{m_j}$ *be linearly independent in the set* X_j*, for each* $j = 1, 2, \ldots, r$*. Set* $M = m_1 m_2 \ldots m_r$*. Then the* M*-tuple of functions*

$$\{f_{i_1}^1(x_1) \cdot f_{i_2}^2(x_2) \cdot \ldots \cdot f_{i_k}^k(x_k) \mid i_j = 1, \ldots, m_j,\ j = 1, \ldots, r\} \tag{3.1.5}$$

is linearly independent in the set $X_1 \times X_2 \times \ldots \times X_r$.

Proof: If the n-tuple $\{f_i\}_{i=1}^{n}$ is linearly independent in X, then Theorem 1.4.1 ensures the existence of elements $u_1, u_2, \ldots, u_n \in X$ satisfying $\det[f_i(u_j)]_{ij} \neq 0$. Then we can apply Cramer's rule to the system

$$h(u_j, y) = \sum_{i=1}^{n} f_i(u_j)g_i(y) \quad (j = 1, 2, \ldots, n)$$

to compute the "unknowns" $g_1, \ldots, g_n$. This proves (i).

To verify (ii) suppose that some linear combination

$$\sum_{i_1=1}^{m_1} \cdots \sum_{i_r=1}^{m_r} \alpha_{i_1 \ldots i_r} f_{i_1}^1(x_1) \cdot \ldots \cdot f_{i_r}^r(x_r) \tag{3.1.6}$$

vanishes identically on $X_1 \times X_2 \times \ldots \times X_r$. For each fixed $x_2 \in X_2, \ldots, x_r \in X_r$, the sum in (3.1.6) is such a linear combination of the m_1-tuple $\{f_i^1\}_{i=1}^{m_1}$ that vanishes on X_1. Since this m_1-tuple is assumed to be linearly independent, we conclude that the equality

$$\sum_{i_2=1}^{m_2} \cdots \sum_{i_r=1}^{m_r} \alpha_{i_1 \ldots i_r} f_{i_2}^2(x_2) \cdot \ldots \cdot f_{i_r}^r(x_r) = 0$$

holds identically on $X_2 \times X_3 \times \ldots \times X_r$, for each $i_1 = 1, 2, \ldots, m_1$. Repeating these arguments $(r-1)$ times, we obtain the identity

$$\sum_{i_r=1}^{m_r} \alpha_{i_1 \ldots i_r} f_{i_r}^r(x_r) = 0 \quad (x_r \in X_r)$$

for each $(r-1)$-tuple of indices $i_1, \ldots, i_{r-1}$. This is possible only in the case when all the coefficients $\alpha_{i_1 \ldots i_r}$ are zero, because of the linear independence of the m_r-tuple $\{f_i^r\}_{i=1}^{m_r}$. □

Proof of Theorem 3.1.1: In view of the induction principle, it is sufficient to prove the following conclusion for $r = 1, 2, \ldots, k-1$:
If the function h from the statement of Theorem 3.1.1 has a representation

$$h(x_1, \ldots, x_k) = \sum_{i_1=1}^{m_1} \cdots \sum_{i_r=1}^{m_r} f_{i_1}^1(x_1) \cdot \ldots \cdot f_{i_r}^r(x_r) \varphi_{i_1 \ldots i_r}^r(x_{r+1}, \ldots, x_k)\,, \quad (3.1.7)$$

then h has also a representation

$$h(x_1, \ldots, x_k) = \sum_{i_1=1}^{m_1} \cdots \sum_{i_{r+1}=1}^{m_{r+1}} f_{i_1}^1(x_1) \cdot \ldots \cdot f_{i_{r+1}}^{r+1}(x_{r+1}) \varphi_{i_1 \ldots i_{r+1}}^{r+1}(x_{r+2}, \ldots, x_k) \quad (3.1.8)$$

(both representations are considered on the whole domain of definition, the Cartesian product $X_1 \times X_2 \times \ldots \times X_k$). Thus assume (3.1.7) to be valid for some r and put $M = m_1 m_2 \ldots m_r$. The hypotheses of Theorem 3.1.1. and Lemma 3.1.2 ensure the existence of M elements

$$(u_1^s, u_2^s, \ldots, u_r^s) \in X_1 \times X_2 \times \ldots \times X_r \quad (s = 1, 2, \ldots, M)$$

such that each function $\varphi_{i_1 \ldots i_r}^r$ from the right-hand side of (3.1.7) is a linear combination of the system of functions

$$\{h(u_1^s, u_2^s, \ldots, u_r^s, -, \ldots, -) \mid s = 1, 2, \ldots, M\} \quad (3.1.9)$$

in the set $X_{r+1} \times \ldots \times X_k$. However, (3.1.1) with $j = r+1$ implies that each function in (3.1.9) (and consequently, each function $\varphi_{i_1 \ldots i_r}^r$ in (3.1.7)) is (in the variable $x_{r+1} \in X_{r+1}$) a linear combination of the m_{r+1}-tuple $\{f_i^{r+1}\}_{i=1}^{m_{r+1}}$, with some coefficients depending on the values of $x_{r+2}, \ldots, x_k$. This immediately leads to a representation (3.1.8). □

Let us repeat the result of our consideration in this section: *A function of $k \geq 3$ variables has a decomposition* (3c) *if and only if there exist k decompositions* (3.1.1) *in two variables* (3.1.3). Since each decomposition (3.1.1) is of type (2a), we may apply a part of results of Chapter 2 here, especially Theorem 2.2.1 on the Casorati determinants. With this application we are led to the following Section 3.2. Note that the variable y in (3.1.3) is multidimensional, which precludes the construction of the Wronskians from Section 2.2 for our decomposition problems (3.1.1). This fact will motivate us to develop a theory of generalized Wronskians in Chapter 3.

Nevertheless, some results on decompositions (3c) can be derived even by using original (non-generalized) Wronskians; we discuss them in Section 3.3.

3.2. Minimal and diagonal decompositions

The method of reduction, described in the preceding section, enables us to state a criterion for decompositions (3c) in terms of the Casorati determinants defined in Section 2.2. To avoid complicated index notations, we restrict the discussion to the case of functions of three variables. For each function $h : X \times Y \times Z \to \mathbb{K}$, we define a triple of the n-th order Casorati determinants

$$\det C_n^x h = \det\big[h(x_i, y_j, z_j)\big]_{i,j=1}^n, \quad \det C_n^y h = \det\big[h(x_j, y_i, z_j)\big]_{i,j=1}^n,$$
$$\det C_n^z h = \det\big[h(x_j, y_j, z_i)\big]_{i,j=1}^n, \quad n = 1, 2, 3, \dots$$

Let us emphasize that each of the determinants $\det C_n^x h$, $\det C_n^y h$ and $\det C_n^z h$ is a scalar function defined on the set $X^n \times Y^n \times Z^n$.

Theorem 3.2.1 [ČŠ 1]: *Let $h \colon X \times Y \times Z \to \mathbb{K}$ be a mapping where X, Y and Z are arbitrary nonempty sets.*

(i) *If h is of the form* (3b) *on the set $X \times Y \times Z$, for some integers m, n and p, then the identities*

$$\det C_{m+1}^x h = 0, \quad \det C_{n+1}^y h = 0 \quad \text{and} \quad \det C_{p+1}^z h = 0 \tag{3.2.1}$$

as well as

$$\det C_{np+1}^x h = 0, \quad \det C_{mp+1}^y h = 0 \quad \text{and} \quad \det C_{mn+1}^z h = 0 \tag{3.2.2}$$

hold on the corresponding domains of definition.

(ii) *Suppose that h has at least one nonzero value and that the identities* (3.2.1) *are valid for some integer $m \geq 1$, $n \geq 1$ and $p \geq 1$. If m, n, p are the smallest numbers satisfying* (3.2.1), *then there exist functions $e_i : X \to \mathbb{K}$, $f_j : Y \to \mathbb{K}$, $g_k : Z \to \mathbb{K}$ and constants $\alpha_{ijk} \in \mathbb{K}$ such that h is of the form* (3b) *on the set $X \times Y \times Z$. Moreover, the numbers m, n and p satisfy the inequalities*

$$m \leq np, \quad n \leq mp \quad \text{and} \quad p \leq mn. \tag{3.2.3}$$

Proof: (i) If h is of the form (3b), then the decompositions

$$h = \sum_{i=1}^m e_i \Big(\sum_{j=1}^n \sum_{k=1}^p \alpha_{ijk} f_j g_k \Big) = \sum_{j=1}^n \sum_{k=1}^p f_j g_k \Big(\sum_{i=1}^m \alpha_{ijk} e_i \Big)$$

imply the first identities in (3.2.1) and (3.2.2), because of the first part of Theorem 2.2.1. The proof of the other identities in (3.2.1) and (3.2.2) is analogous.

(ii) If m, n, p are the smallest integers satisfying (3.2.1), then the second part of Theorem 2.2.1 yields decompositions

$$\begin{aligned} h(x,y,z) &= \sum_{i=1}^{m} e_i(x)\varphi_i(y,z) = \sum_{j=1}^{n} f_j(y)\psi_j(x,z) \\ &= \sum_{k=1}^{p} g_k(z)\tau_k(x,y) \end{aligned} \tag{3.2.4}$$

on the set $X \times Y \times Z$, with the groups of functions $\{e_i\}$, $\{f_j\}$ and $\{g_k\}$, being linearly independent in the sets X, Y and Z respectively. Consequently, the existence of a decomposition (3b) is ensured by Theorem 3.1.1 with $k=3$. Finally, inequalities (3.2.3) follow from the minimality of the numbers m, n, p and from the identities (3.2.2), proved in part (i). □

Let us turn now our attention to the following question. Which of all the possible decompositions (3c) of a given function h should be called *minimal*? Recalling the definition of minimal decompositions for functions of two variables (Section 2.3), one could suppose that all we need to assume for (3c) to be minimal is the linear independence of the m_j-tuple $\{f_i^j\}_{i=1}^{m_j}$ in the set X_j, for each $j = 1, 2, \ldots, r$. However, the role of the coefficients $\alpha_{i_1 \ldots i_k}$ is neglected in such a definition: it is clear that a kind of *dependence* of the coefficients $\alpha_{i_1 \ldots i_k}$ enables us to *reduce* the representation (3c). For example, such a reduction is possible in the case when an m_j-tuple $\{\varphi_i^j\}_{i=1}^{m_j}$ in some of k decompositions (3.1.1) is linearly dependent. Thus *we call a decomposition* (3c) *to be minimal* if and only if each of the k induced decompositions (3.1.1) is minimal in the sense of Section 2.3 (with respect to the pair of variables (3.1.3)). In view of part (ii) of Lemma 3.1.2, we can state an explicit criterion for a decomposition (3c) to be minimal without referring to the minimality of (3.1.1).

Lemma 3.2.2: *Let* $h\colon X_1 \times X_2 \times \ldots \times X_k \to \mathbb{K}$ *be a function, where* X_1, X_2, ..., X_k *are arbitrary nonempty sets. Then a decomposition* (3c) *is minimal if and only if the following pair of conditions is fulfilled for each* $j = 1, 2, \ldots, k$*:*

(i) *the* m_j*-tuple of functions* $\{f_i^j\}_{i=1}^{m_j}$ *is linearly independent in the set* X_j*;*

(ii) *the system of* $m_1 \cdots m_{j-1} \cdot m_{j+1} \cdots m_k$ *linear algebraic equations*

$$\sum_{i=1}^{m_j} \alpha_{i_1 \ldots i_{j-1}\, i\, i_{j+1} \ldots i_k} \beta_i = 0 \quad (i_r = 1, \ldots, m_r;\ r = 1, \ldots, j-1, j+1, \ldots, k)$$

has the unique solution $\beta_1 = \beta_2 = \ldots = \beta_{m_j} = 0$.

Proof: Since (i) is necessary for (3c) to be minimal by definition, we have to verify the following conclusion: *Suppose that* (i) *holds for each* $j = 1, \ldots, k$. *Then the* m_j-*tuple* $\{\varphi_i^j\}_{i=1}^{m_j}$ *from* (3.1.1) *is linearly independent* (in the set Y_j defined in Theorem 3.1.1) *if and only if the condition* (ii) *holds.* In fact, each function φ_i^j equals the sum of all the terms (3.1.2) and, by part (ii) of Lemma 3.1.2, the system of functions

$$\{f_{i_1}^1 \cdot \ldots \cdot f_{i_{j-1}}^{j-1} \cdot f_{i_{j+1}}^{j+1} \cdot \ldots f_{i_k}^k \mid i_r = 1, \ldots, m_r;\ r = 1, \ldots, j-1, j+1, \ldots, k\}$$

is linearly independent in Y_j. Thus a linear combination $\sum_{i=1}^{m_j} \beta_i \varphi_i^j$ vanishes on Y_j if and only if the m_j-tuple of constants β_i solves the system of equations introduced in the statement of condition (ii). □

Note that the *existence* of minimal decompositions, for each function h of type (3c), is ensured by Theorem 3.1.1: starting from k minimal decompositions (3.1.1), we necessarily obtain a minimal decomposition (3c). The main properties of the minimal decompositions are described in the following generalization of Theorem 2.3.1.

Theorem 3.2.3: *Let*

$$h(x_1, \ldots, x_k) = \sum_{i_1=1}^{m_1} \cdots \sum_{i_k=1}^{m_k} \alpha_{i_1 \ldots i_k} f_{i_1}^1(x_1) \cdot \ldots \cdot f_{i_k}^k(x_k) \qquad (3.2.5a)$$

$$h(x_1, \ldots, x_k) = \sum_{i_1=1}^{\tilde{m}_1} \cdots \sum_{i_k=1}^{\tilde{m}_k} \tilde{\alpha}_{i_1 \ldots i_k} \tilde{f}_{i_1}^1(x_1) \cdot \ldots \cdot \tilde{f}_{i_k}^k(x_k) \qquad (3.2.5b)$$

be two arbitrary decompositions of the same function $h\colon X_1 \times X_2 \times \ldots \times X_k \to \mathbb{K}$ *and let* (3.2.5a) *be a minimal one. Then*

$$\tilde{m}_j \geq m_j \qquad (j = 1, 2, \ldots, k) \qquad (3.2.6)$$

and (3.2.5b) *is minimal if and only if all the equalities in* (3.2.6) *hold. If this is the case, then there exist* k *constant nonsingular matrices* $C^j = [\gamma_{ik}^j]_{i,k=1}^{m_j}$ *such that*

$$\begin{pmatrix} \tilde{f}_1(x) \\ \tilde{f}_2(x) \\ \vdots \\ \tilde{f}_{m_j}(x) \end{pmatrix} = (C^j)^{-1} \cdot \begin{pmatrix} f_1(x) \\ f_2(x) \\ \vdots \\ f_{m_j}(x) \end{pmatrix} \qquad (x \in X_j,\ j = 1, 2, \ldots, k) \qquad (3.2.7)$$

and the equality

$$\tilde{\alpha}_{i_1 i_2 \dots i_k} = \sum_{j_1=1}^{m_1} \sum_{j_2=1}^{m_2} \cdots \sum_{j_k=1}^{m_k} \gamma^1_{i_1 j_1} \gamma^2_{i_2 j_2} \dots \gamma^k_{i_k j_k} \cdot \alpha_{j_1 j_2 \dots j_k} \tag{3.2.8}$$

holds for each k-tuple of indices $(i_1, \dots, i_k)$.

Proof: Applying Theorem 2.3.1 to each pair of decompositions

$$\sum_{i=1}^{m_j} f_i^j(x_j)\varphi_i^j(x_1, \dots, x_{j-1}, x_{j+1}, \dots, x_k)$$
$$= \sum_{i=1}^{\tilde{m}_j} \tilde{f}_i^j(x_j)\tilde{\varphi}_i^j(x_1, \dots, x_{j-1}, x_{j+1}, \dots, x_k)$$

originated from (3.2.5a) and (3.2.5b) respectively, we obtain system (3.2.6) and the criterion of k equalities in it. Moreover, if $\tilde{m}_j = m_j$ for each $j = 1, 2, \dots, k$, Theorem 2.3.1 also ensures the existence of k matrices C^j satisfying (3.2.7). Multiplying (3.2.7) from the left by C^j and substituting the result into (3.2.5a), we get another decomposition

$$h(x_1, \dots, x_k) = \sum_{i_1=1}^{\tilde{m}_1} \cdots \sum_{i_k=1}^{\tilde{m}_k} \hat{\alpha}_{i_1 \dots i_k} \tilde{f}^1_{i_1}(x_1) \cdot \ldots \cdot \tilde{f}^k_{i_k}(x_k) \tag{3.2.9}$$

in which each coefficient $\hat{\alpha}_{i_1 \dots i_k}$ is equal to the right-hand side of (3.2.8). By virtue of part (ii) of Lemma 3.1.2, the system of functions

$$\left\{ \tilde{f}^1_{i_1}(x_1) \cdot \tilde{f}^2_{i_2}(x_2) \cdot \ldots \cdot \tilde{f}^k_{i_k}(x_k) \mid i_j = 1, \dots, m_j;\ j = 1, \dots, k \right\}$$

is linearly independent in $X_1 \times \ldots \times X_k$. Thus the comparison of (3.2.5b) and (3.2.9) leads to the conclusion that $\tilde{\alpha}_{i_1 \dots i_k} = \hat{\alpha}_{i_1 \dots i_k}$, for each k-tuple of indices $(i_1, \dots, i_k)$, which proves (3.2.8). □

Remarks 3.2.4: **(i)** The parameters $m_1, \dots, m_k$ of minimal decompositions (3c) can be determined by means of the Casorati determinants as the smallest integers satisfying k identities analogous to (3.2.1) for the case $k = 3$. It follows from condition (ii) of Lemma 3.2.2 that the parameters $m_1, \dots, m_k$ have to satisfy the k inequalities

$$m_j \le m_1 \cdots m_{j-1} m_{j+1} \cdots m_k \qquad (j = 1, 2, \dots, k)$$

(for the case $k = 3$, see also (3.2.3)).

(ii) The components f_i^j of each minimal decomposition (3c) can be expressed as linear combinations of the suitable m_j-tuple of "partial" functions

$$\{h(x_1^{ij},\ldots,x_{j-1}^{ij},-,x_{j+1}^{ij},\ldots,x_k^{ij}) \mid i=1,2,\ldots,m_j\}.$$

This fact follows from part (ii) of Corollary 2.3.2, applied to (3.1.1). Hence it is possible to describe all the minimal decompositions of a given function h by a formula analogous to (2.3.13). We omit this simple discussion here because it necessitates the introduction of a rather complicated index notation.

Now we shall deal with diagonal decompositions (3a). The main result in this direction, stated here as Theorem 3.2.5, can be briefly read as follows: If h has a decomposition (3a), with a fixed number of terms $N \geq 1$, then there exists also such a decomposition (3a) (with the same N), in which the components e_i, f_i, g_i are linear combinations of the corresponding components, taken from any minimal decomposition (3b).

Theorem 3.2.5 [ČŠ 1]: *Let $h\colon X \times Y \times Z \to \mathbb{K}$ be a function, where X, Y and Z are arbitrary nonempty sets. If h is of the form*

$$h(x,y,z) = \sum_{s=1}^{N} \hat{e}_s(x)\hat{f}_s(y)\hat{g}_s(z) \quad (x \in X,\ y \in Y,\ z \in Z) \tag{3.2.10}$$

for some integer $N \geq 1$ and if (3b) *is a minimal decomposition of h on the set $X \times Y \times Z$, then there exist constants $\lambda_{si}, \mu_{sj}, \nu_{sk} \in \mathbb{K}$ such that the $3N$ functions*

$$\tilde{e}_s := \sum_{i=1}^{m} \lambda_{si} e_i, \quad \tilde{f}_s := \sum_{j=1}^{n} \mu_{sj} f_j, \quad \tilde{g}_s := \sum_{k=1}^{p} \nu_{sk} g_k \quad (s = 1,\ldots,N) \tag{3.2.11}$$

satisfy the identity

$$h(x,y,z) = \sum_{i=1}^{N} \tilde{e}_i(x)\tilde{f}_i(y)\tilde{g}_i(z) \quad (x \in X,\ y \in Y,\ z \in Z). \tag{3.2.12}$$

The constants $\lambda_{si}, \mu_{sj}, \nu_{sk}$ possess this property if and only if they solve the system of mnp equations

$$\sum_{s=1}^{N} \lambda_{si}\mu_{sj}\nu_{sk} = \alpha_{ijk} \tag{3.2.13}$$

where α_{ijk} are the mnp coefficients taken from (3b).

Proof: Let U and V be linear spaces generated by the groups of functions $\{\hat{e}_s\}_{s=1}^N$ from (3.2.10) and $\{e_i\}_{i=1}^m$ from (3b), respectively. Since (3b) is assumed to be minimal, the m-tuple functions $\{\varphi_i\}_{i=1}^m$ from the induced decomposition (3.2.4) is linearly independent in the set $Y \times Z$. Consequently, part (i) of Lemma 3.1.2 implies that each function e_i is a linear combination of some m-tuple of functions $\{h(-, y_j, z_j)\}_{j=1}^m$, which means that $V \subseteq U$. Let $P : U \to V$ be any linear projection onto V. Then for each pair $(y, z) \in Y \times Z$, we conclude from (3.2.4) and (3.2.11) that

$$\sum_{s=1}^{N} \hat{f}_s(y)\hat{g}_s(z)P\hat{e}_s = P\Big(\sum_{s=1}^{N} \hat{f}_s(y)\hat{g}_s(z)\hat{e}_s\Big) = P(h(-,y,z))$$
$$= \sum_{i=1}^{m} \varphi_i(y,z)Pe_i = \sum_{i=1}^{m} \varphi_i(y,z)e_i = h(-,y,z)\,.$$

Setting $\tilde{e}_s = P\hat{e}_s$ for each s, we therefore obtain

$$\tilde{e}_s = \sum_{i=1}^{m} \lambda_{si}e_i \quad \text{and} \quad h(x,y,z) = \sum_{s=1}^{N} \tilde{e}_s(x)\hat{f}_s(y)\hat{g}_s(z) \quad \text{on } X \times Y \times Z\,.$$

Repeating this projective procedure in the variables y and z, we get at the end a decomposition (3.2.12), with components of the form (3.2.11). Recall that the system $\{e_i \cdot f_j \cdot g_k\}_{i=1}^{m}{}_{j=1}^{n}{}_{k=1}^{p}$ is linearly independent on $X \times Y \times Z$, because of part (ii) of Lemma 3.1.2. Thus substituting (3.2.11) into (3.2.12) and comparing the result with (3b), we obtain equations (3.2.13). Conversely, if the numbers λ_{si}, μ_{sj}, ν_{sk} satisfy the system (3.2.13) and if the functions $\tilde{e}_s$, $\tilde{f}_s$, $\tilde{g}_s$ are defined by (3.2.11), then an easy computation shows that the right-hand sides of (3.2.12) and (3b) are the same, i.e. equal to the function h. □

Remark 3.2.6: Theorem 3.2.5 provides a procedure for finding a diagonal decomposition (3a) of the given function h, with a minimal number N of terms: using Theorem 3.2.1 and Remark 3.2.4(ii), we first determine a minimal decomposition (3b). Taking then the coefficients α_{ijk} from (3b), we construct the system of mnp equations (3.2.13). Finally, we seek the smallest N, for which this system is solvable (we are not in a position to solve this purely algebraic question in general). Let us note only that the unknown smallest N has clearly to satisfy the estimates

$$\max\{m,n,p\} \le N \le mnp\,.$$

3.3. Differential criteria

In this section we prove some necessary and some sufficient conditions for a given smooth function h of k real variables $x_1, x_2, \ldots, x_k$ to be of the form (3c). Let us emphasize that a general approach to this problem, based on the method of reduction

(Section 3.1) and the theory of generalized Wronskians (Chapter 4), will be described in Chapter 5.

From the historical point of view, the first result [N 3] concerning decompositions (3c) can be essentially stated in terms of the following theorem. (In fact, the original statement in [N 3] concerns the case when $m_1 = \ldots = m_k$.)

Theorem 3.3.1 [N 3]: *Let $I_1, I_2, \ldots, I_k$ be $k \geq 3$ intervals in $\mathbb{R}$ and let m_1, m_2, $\ldots$, m_k be positive integers. Suppose that a function $h\colon I_1 \times \ldots \times I_k \to \mathbb{K}$ has the partial derivative $h_{x_1^{m_1} x_2^{m_2} \ldots x_k^{m_k}}$, which is continuous in all variables on $I_1 \times \ldots \times I_k$. Then the function h can be written in the form* (3c)*, in which the Wronskian of each m_j-tuple $\{f_i^j\}_{i=1}^{m_j}$ has no zero value in the whole interval I_j ($j = 1, 2, \ldots, k$), if and only if for each $j = 1, 2, \ldots, k$, the function h (in the variable x_j) is a solution of the (unique) linear ordinary differential equation of order m_j*

$$y^{(m_j)} + p_1^j(x) y^{(m_j - 1)} + \cdots + p_{m_j}^j(x) y = 0 \quad (y = y(x),\ x \in I_j) \tag{3.3.1}$$

on the whole interval I_j. The coefficients $p_i^j : I_j \to \mathbb{K}$ in (3.3.1) *do not depend on the choice of the other variables $x_1, \ldots, x_{j-1}, x_{j+1}, \ldots, x_k$.*

Proof: By the method of reduction from Section 3.1, a decomposition (3c) of h exists if and only if for each $j = 1, 2, \ldots, k$, h can be written in the form

$$h(x_1, \ldots, x_k) = \sum_{i=1}^{m_j} f_i^j(x_j) \varphi_i^j(x_1, \ldots, x_{j-1}, x_{j+1}, \ldots, x_k)\,. \tag{3.3.2}$$

If the Wronskian of the m_j-tuple $\{f_i^j\}_{i=1}^{m_j}$ has no zero value on I_j, then this m_j-tuple forms a fundamental set of solutions of the (unique) equation (3.3.1). Since the function h in (3.3.2) is (in the variable x_j) a linear combination of this m_j-tuple of solutions, h is a solution of (3.3.1), too. Conversely, suppose that there exist k equations (3.3.1) and that h solves each of them in the corresponding variable. Choose a fundamental set of solutions of (3.3.1) and denote them as $\{f_i^j\}_{i=1}^{m_j}$. Then the function h, another solution of (3.3.1), is (in the variable x_j) a linear combination of the m_j-tuple $\{f_i^j\}_{i=1}^{m_j}$, with some coefficients that may depend on the other variables $x_1, \ldots, x_{j-1}, x_{j+1}, \ldots, x_k$. This is exactly (3.3.2) and the proof is complete. □

Remark 3.3.2: The proof of Theorem 3.3.1 is non-constructive: for a given function h of type (3c), with unknown components f_i^j, the method of the proof gives no answer to the question of finding the system of k ordinary differential equations (3.3.1).

Partial Wronskians: It is clear that a certain kind of information about decompositions (3c) can be obtained by the following construction. Given a smooth function h of $k \geq 3$ real variables $x_1, \ldots, x_k$, we can consider the sequence of the Wronski matrices of the type (2.2.1)

$$W_n^{x_i,x_j}h := \begin{pmatrix} h & h_{x_j} & \dots & h_{x_j^{n-1}} \\ h_{x_i} & h_{x_i x_j} & \dots & h_{x_i x_j^{n-1}} \\ \vdots & \vdots & \ddots & \vdots \\ h_{x_i^{n-1}} & h_{x^{n-1} x_j} & \dots & h_{x_i^{n-1} x_j^{n-1}} \end{pmatrix} \qquad (n = 1, 2, \dots) \tag{3.3.3}$$

for each pair of different indices $i, j \in \{1, 2, \dots, k\}$. It is easy to understand why we call $\det W_n^{x_i,x_j} h$ the *partial Wronski determinant* (or the *partial Wronskian*) of the function h. By virtue of Theorem 2.2.1, we can immediately state a necessary condition for a smooth function $h\colon I_1 \times \dots \times I_k \to \mathbb{K}$ to be of the form (3c): the identity

$$\det W_{n+1}^{x_i,x_j} h = 0 \qquad \text{(on the set } I_1 \times \dots \times I_k) \tag{3.3.4}$$

is valid for each $n \geq \min\{m_i, m_j\}$, $1 \leq i < j \leq k$. This motivates us to solve the "converse" problem: What form must a function h be of, provided that identity (3.3.4) holds with some integer $n = n_{ij}$, for each pair of different indices $i, j \in \{1, 2, \dots, k\}$? Let us give two examples showing that the system of $\binom{k}{2}$ identities (3.3.4) is not very suitable for finding decompositions (3c) in the following sense. *In general, the conditions* (3.3.4) *cannot determine the parameters* $m_1, \dots, m_k$ *of the minimal decompositions* (3c). As the first example, consider a function possessing the diagonal decomposition

$$h(x_1, \dots, x_k) = \sum_{i=1}^{N} f_i^1(x_1) \cdot \ldots \cdot f_i^k(x_k)$$

for which the smallest integer $n = n_{ij}$ in (2.3.4) is equal to the same number $n_{ij} = N$ (in general). The function

$$h(x_1, \dots, x_k) = \prod_{1 \leq i < j \leq k} \left(\sum_{s=1}^{n_{ij}} f_{ijs}(x_i) g_{ijs}(x_j) \right)$$

may serve as a converse example: While such a function is of the form (3c), with the "large" parameters

$$m_j = n_{1j} n_{2j} \dots n_{j-1,j} n_{j,j+1} n_{j,j+2} \dots n_{jk}$$

(minimal in general), the identities (3.3.4) are valid for $n = n_{ij}$, because h in the variables x_i, x_j $(i \neq j)$ can be written as

$$h = \sum_{s=1}^{n_{ij}} \big(A(x_i) f_{ijs}(x_i)\big) \cdot \big(B(x_j) g_{ijs}(x_j)\big) .$$

Nevertheless, in the case $k = 3$, the following assertion on the conditions (3.3.4) can be proved.

Theorem 3.3.3 [ČŠ 1]: *Let $h\colon I \times J \times K \to \mathbb{K}$ be a smooth function, where I, J, K are three intervals in $\mathbb{R}$. Suppose that the partial Wronskians of the function h satisfy the conditions*

$$\det W_{M+1}^{x,y} h = 0, \quad \det W_{N+1}^{y,z} h = 0, \quad \det W_{P+1}^{x,z} h = 0 \tag{3.3.5}$$

and

$$\det W_{M}^{x,y} h \neq 0, \quad \det W_{N}^{y,z} h \neq 0, \quad \det W_{P}^{x,z} h \neq 0 \tag{3.3.6}$$

on the whole set $I \times J \times K$, for some integers $M,\ N,\ P \geq 1$. Then the function h has a decomposition (3b) *on $I \times J \times K$ such that the parameters m, n, p in it satisfy the estimates*

$$m \leq MP, \quad n \leq MN \quad \text{and} \quad p \leq NP. \tag{3.3.7}$$

Proof: Having successively fixed each of the variables x, y, z, we observe that Theorem 2.2.1 and conditions (3.3.5) and (3.3.6) ensure the existence of three decompositions

$$\begin{aligned} h(x,y,z) &= \sum_{i=1}^{M} \hat{e}_i(x,z)\hat{\varphi}(y,z) = \sum_{j=1}^{N} \hat{f}_j(x,y)\hat{\psi}_j(x,z) \\ &= \sum_{k=1}^{P} \hat{g}_k(y,z)\hat{\tau}_k(x,y) \end{aligned} \tag{3.3.8}$$

on the set $I \times J \times K$. Moreover, the Wronskian of the M-tuple of functions $\{\hat{\varphi}_i\}$ in the variable y satisfies (see Remark 2.1.2(ii))

$$\det W(\hat{\varphi}_1(-,z), \ldots, \hat{\varphi}_M(-,z); y) \neq 0 \quad (y \in J, z \in K). \tag{3.3.9}$$

Differentiating the first equality in (3.3.8) repeatedly with respect to y, we obtain the following system of M equalities

$$\sum_{i=1}^{M} \hat{e}_i(x,z) \frac{\partial^r \hat{\varphi}}{\partial y^r}(y,z) = \frac{\partial^r}{\partial y^r}\left(\sum_{k=1}^{P} \hat{g}_k(y,z)\hat{\tau}_k(x,y) \right) \quad (r = 0, 1, \ldots, M-1)\,.$$

In view of (3.3.9), we can apply Cramer's rule to this system with a fixed $y = y_0$ to compute the unknown functions $\hat{e}_i$. The result of this application can be written in the form

$$\hat{e}_i(x,z) = \sum_{k=1}^{P}\sum_{r=0}^{M-1} \xi_{ikr}(z)\frac{\partial^r \hat{\tau}_k}{\partial y^r}(x,y_0)$$
$$= \sum_{s=1}^{m} e_s(x)\zeta_{is}(z) \qquad (x\in I,\ z\in K,\ i=1,\dots,M)$$

where the m-tuple $e_1,\dots,e_m$ is any basis of the linear space generated by the following system of MP functions

$$\left\{ \frac{\partial^r \hat{\tau}_k}{\partial y^r}(-,y_0) \mid r=0,1,\dots,M-1;\ k=1,2,\dots,P \right\} \tag{3.3.10}$$

considered on the interval I. Thus $m \leq MP$ and the substitution (3.3.10) into (3.3.8) yields a decomposition

$$h(x,y,z) = \sum_{i=1}^{M} \hat{e}_i(x)\hat{\varphi}(y,z) = \sum_{i=1}^{M}\Big(\sum_{s=1}^{m} e_s(x)\zeta_{is}(z)\Big)\hat{\varphi}(y,z)$$
$$= \sum_{i=1}^{m} e_i(x)\varphi_i(y,z)$$

on the set $I \times J \times K$. Analogously we derive the other two decompositions in (3.2.4), in which $n \leq MN$ and $p \leq NP$. In view of Theorem 3.1.1 with $k=3$, the decompositions (3.2.4) ensure the existence of a decomposition (3b), with parameters m, n and p satisfying (3.3.7). □

Remark 3.3.4: The restriction of the functions h to be of three variables only is very *essential* in the above proof. Thus it seems to be an *open problem* to extend Theorem 3.3.3 to the case of functions of more than three variables.

Let us finish this section with another procedure of finding decompositions (3c) of functions of $k \geq 3$ real variables, based on the method of reduction described in Section 3.1. Namely, each of the k needed decompositions (3.1.1) is of the form

$$h(x,y,z) = \sum_{i=1}^{n} f_i(x)g_i(y,z) \tag{3.3.11}$$

where x, y are real variables and z is a $(k\text{–}2)$-dimensional vector variable, considered to be a parameter. Thus a criterion for h to be of type (3.3.11) can be obtained in the following way. After utilizing the partial Wronskian of h in the real variables x, y to obtain a decomposition

$$h(x,y,z) = \sum_{i=1}^{n} \hat{f}_i(x,z)\hat{g}_i(y,z)\,, \qquad (3.3.12)$$

we have to ensure that the components $\hat{f}_i$ in the right-hand side of the (minimal) decomposition (3.3.12) can be chosen to be independent of the parameter z. By Theorem 2.3.1, this choice is possible if and only if there exists a nonsingular matrix $C = C(z)$ of size $n \times n$ such that any of the following n-tuple of functions

$$\begin{pmatrix} f_1(x) \\ f_2(x) \\ \vdots \\ f_n(x) \end{pmatrix} = C^T(z) \cdot \begin{pmatrix} \hat{f}_1(x,z) \\ \hat{f}_2(x,z) \\ \vdots \\ \hat{f}_n(x,z) \end{pmatrix} \qquad (3.3.13)$$

does not depend on the variable z. This existence condition is obviously fulfilled in the case when the n-tuples $\{f_i\}$ and $\{\hat{f}_i\}$ are fundamental sets of solutions of the same linear ordinary differential equation of order n (in the variable x). This idea underlies the following:

Theorem 3.3.5 [ČŠ 1]: *Let $h : I \times J \times Z \to \mathbb{K}$ be a function, where I, J are intervals in $\mathbb{R}$ and Z is an arbitrary nonempty set. Suppose that for each $z \in Z$, the function h has the partial derivative $h_{x^n y^n}$, which is continuous on $I \times J$.*

(i) *If the function h has a decomposition* (3.3.11) *on $I \times J \times Z$, then the identity* $\det W_{n+1}^{x,y} h = 0$ *holds on $I \times J \times Z$. If in addition the inequality* $\det W_n^{x,y} h \neq 0$ *holds on $I \times J \times Z$, then the coefficients of the n-th-order linear differential equation*

$$\frac{1}{\det W_n^{x,y} h} \cdot \det \begin{pmatrix} h & h_y & \dots & h_{y^{n-1}} & f \\ h_x & h_{xy} & \dots & h_{xy^{n-1}} & f' \\ \vdots & \vdots & \ddots & \vdots & \vdots \\ h_{x^n} & h_{x^n y} & \dots & h_{x^n y^{n-1}} & f^{(n)} \end{pmatrix} = 0\,, \qquad (3.3.14)$$

in which $f = f(x)$ is the "unknown" function, do not depend on $y \in J$ and $z \in Z$.

(ii) *If h satisfies* $\det W_{n+1}^{x,y} h = 0$ *and* $\det W_n^{x,y} h \neq 0$ *on the set $I \times J \times Z$ and if there exists $y_0 \in J$ such that the coefficients of equation* (3.3.14) *with $y = y_0$ do not depend on $z \in Z$, then h has a decomposition* (3.3.11) *on $I \times J \times Z$, in which the n-tuple $\{f_i\}$ can be chosen as any fundamental set of solutions of* (3.3.14).

Proof: (i) The identity $\det W_{n+1}^{x,y} h = 0$ follows from Theorem 2.1.1. Notice that the differential equation (3.3.14) has a fundamental set of solutions

$$h(-,y,z), h_y(-,y,z), \dots, h_{y^{n-1}}(-,y,z)$$

for each (fixed) $y \in J$ and $z \in Z$. In view of part (i) of Corollary 2.3.2, this n-tuple of functions forms a basis of the linear space generated by the n-tuple $\{f_i\}$ from a minimal decomposition (3.3.11). Consequently, the space of all solutions of (3.3.14) does not depend on $y \in Y$ and $z \in Z$; hence the coefficients of (3.3.14) do not in turn depend on $y \in Y$ and $z \in Z$.

(ii) If h satisfies the conditions stated in part (ii), then Theorem 2.1.1 forces a decomposition (3.3.12). Moreover, Theorem 2.1.5 implies that the components $\hat{f}_i$ in (3.3.12) can be taken in the form

$$\hat{f}_i(x,z) = h_{y^{i-1}}(x, y_0, z) \qquad (i=1,2,\ldots,n)$$

where $y_0 \in J$ is a fixed point from the statement of (ii). Notice that this n-tuple $\{\hat{f}_i\}$ is (in the variable x) a fundamental set of solutions of (3.3.14) with $y = y_0$. Since the coefficients of this equation are assumed to be independent of $z \in Z$, there exists another fundamental n-tuple of solutions $\{f_i\}$, being also independent of $z \in Z$. Since $\{\hat{f}_i\}$ and $\{f_i\}$ are two bases of the same linear space, there exists a nonsingular matrix $C = C(z)$ such that (3.3.13) holds on the whole interval I. Since the substitution (3.3.13) transforms (3.3.12) to a decomposition of type (3.3.11), the proof is complete. □

Remark 3.3.6: Equation (3.3.14) is of the form

$$f^{(n)} + p_1(x,y,z)f^{(n-1)} + \cdots + p_n(x,y,z)f = 0$$

in which the coefficients p_i can be obtained by expanding the determinant in (3.3.14) with respect to the last column. Thus in the statement of Theorem 3.3.5, it would be possible to omit equation (3.3.14) and to replace it by the explicit formulas for the coefficients p_i. However, we believe the above statement is easier and that the reader can compute the coefficients p_i him/herself, if necessary. Let us add the fact that in the case when h is a smooth function in the vector parameter $\vec{z} = (z_1, \ldots, z_k) \in Z$, Z is a region in $\mathbb{R}^k$, then the independence of the coefficient p_i on the paramenter $\vec{z}$ is equivalent to the system of k equalities

$$\frac{\partial p_i}{\partial z_j} = 0 \quad \text{on } Z \quad (j=1,2,\ldots,k).$$

3.4. Matrix factorizations

This section deals mainly with the factorization problem (3d), in which the values of H and F_i are nonsingular matrices of the same size $n \times n$. (Recall that the case $k=2$ was considered in Section 2.4.)

Before we state an extension of Theorem 2.4.1, the basic result of Section 2.4, we introduce the following notation. Given a mapping $H : X_1 \times X_2 \times \ldots \times X_k \to$

$GL_n(\mathbb{K})$ and k chosen elements $u_i \in X_i$, $i = 1, \dots, k$, we define the k-tuple of *partial mappings* $H_i : X_i \to GL_n(\mathbb{K})$ by means of the rule

$$H_i(x) := H(u_1, \dots, u_{i-1}, x, u_{i+1}, \dots, u_k) \quad (x \in X_i,\ 1 \le i \le k)\,. \qquad (3.4.1)$$

Theorem 3.4.1 [Ši 4]: *Let $H : X_1 \times \dots \times X_k \to GL_n(\mathbb{K})$ be a mapping, where $X_1, \dots, X_k$ are $k \ge 2$ nonempty sets. Consider partial mappings* (3.4.1) *for a fixed k-tuple $u_1, \dots, u_k$ and denote $H_0 := H(u_1, \dots, u_k)$. Then the mapping H has a factorization* (3d) *on $X_1 \times \dots \times X_k$ if and only if it satisfies the equation*

$$H(\boldsymbol{x}) = H_1(x_1) \cdot H_0^{-1} \cdot H_2(x_2) \cdot H_0^{-1} \cdot \dots \cdot H_0^{-1} \cdot H_k(x_k) \qquad (3.4.2)$$

for each $\boldsymbol{x} = (x_1, \dots, x_k) \in X_1 \times \dots \times X_k$. Moreover, the factors $F_i : X_i \to GL_n(\mathbb{K})$ from any factorization (3d) *are given by*

$$\begin{aligned} F_1(x) &= H_1(x) \cdot C_1 \\ F_i(x) &= D_{i-1} \cdot H_i(x) \cdot C_i \quad (1 < i < k) \\ F_k(x) &= D_{k-1} \cdot H_k(x) \end{aligned} \qquad (3.4.3)$$

where $C_i, D_i \in GL_n(\mathbb{K})$ are arbitrary constant matrices satisfying

$$C_1 \cdot D_1 = C_2 \cdot D_2 = \dots = C_{k-1} \cdot D_{k-1} = H_0^{-1}. \qquad (3.4.4)$$

Proof: If H is as in (3d), then one can check inductively that

$$H_1(x_1) H_0^{-1} \dots H_0^{-1} H_i(x_i) = F_1(x_1) \dots F_i(x_i) F_{i+1}(u_{i+1}) \dots F_k(u_k),$$

for each $i = 2, 3, \dots, k$. This equality with $i = k$ proves (3.4.2). Conversely, if the mapping H satisfies (3.4.2), then it is clearly of type (3d), for example, with factors $F_1 = H_1$ and $F_i = H_0^{-1} H_i$, $i = 2, 3, \dots, k$ (as well as with factors (3.4.3) under condition (3.4.4)). Consequently, it remains to show that the factors F_i from any factorization (3d) must be of type (3.4.3). Indeed, it follows from (3d) that

$$\begin{aligned} F_1(x) &= H_1(x) H_0^{-1} F_1(u_1) \\ F_i(x) &= F_{i-1}^{-1}(u_{i-1}) F_{i-2}^{-1}(u_{i-2}) \dots F_1(u_1)^{-1} H_i(x) H_0^{-1} F_1(u_1) F_2(u_2) \dots F_i(u_i) \\ &\qquad (i = 2, 3, \dots, k-1) \\ F_k(x) &= F_{k-1}^{-1}(u_{k-1}) \dots F_1(u_1)^{-1} H_k(x)\,. \end{aligned}$$

Thus F_i are of type (3.4.3), with matrices

$$C_i = H_0^{-1} F_1(u_1) F_2(u_2) \dots F_i(u_i) \quad \text{and} \quad D_i = F_i^{-1}(u_i) F_{i-1}^{-1}(u_{i-1}) \dots F_1^{-1}(u_1),$$

which clearly satisfy (3.4.4). □

Comparing functional equation (3.4.2) with equation (2.4.3), one may analogously presume that

$$H_{x_1x_2\ldots x_k} = H_{x_1} \cdot H^{-1} \cdot H_{x_2} \cdot H^{-1} \cdot \ldots \cdot H^{-1} \cdot H_{x_k} \tag{3.4.5}$$

is a *good* generalization of differential equation (2.4.8). We disprove this conjecture below:

Theorem 3.4.2: *Let $I_1, \ldots, I_k$ be $k \geq 2$ intervals in $\mathbb{R}$ and let the mapping $F_j : I_j \to GL_n(\mathbb{K})$ be differentiable at each point of I_j, $j=1,\ldots,k$. Then the mapping H defined by* (3d) *is a solution of* (3.4.5) *on the set $I_1 \times \ldots \times I_k$. However, in the case when $k \geq 3$, equation* (3.4.5) *has such solutions which are not of the form* (3d).

Proof: If H is as in (3d), with differentiable factors F_i, then one can check inductively that

$$H_{x_1}H^{-1}H_{x_2}H^{-1}\ldots H^{-1}H_{x_i}H^{-1} = F_1'F_2'\ldots F_i'F_{i-1}^{-1}F_{i-2}^{-1}\ldots F_1^{-1}$$

for $i=1,2,\ldots,k-1$ and, in the last step,

$$H_{x_1}H^{-1}H_{x_2}H^{-1}\ldots H^{-1}H_{x_k} = F_1'F_2'\ldots F_k' = H_{x_1x_2\ldots x_k}\,.$$

Hence H solves (3.4.5). On the other hand, any smooth mapping H not depending on x_1, i.e. $H=H(x_2,\ldots,x_k)$, is an example of a solution of (3.4.5), which is not of type (3d) in general (provided that $k \geq 3$). □

The above result shows that equation (3.4.5) presents a necessary (but not a sufficient) condition for a smooth mapping H to have a factorization (3d). To obtain a sufficient condition (see Theorem 3.4.4), we develop another reduction principle (Lemma 3.4.3) showing that the factorization problem (3d) is equivalent to a system of $\binom{k}{2}$ problems (2b).

Given a mapping $H : X_1 \times \ldots \times X_k \to GL_n(\mathbb{K})$, let us introduce the families of $\binom{k}{2}$ partial mappings $H_{ij} : X_i \times X_j \to GL_n(\mathbb{K})$, where $1 \leq i < j \leq k$, defined by

$$H_{ij}(x_i, x_j) := H(x_1, \ldots, x_k) \quad (x_i \in X_i,\ x_j \in X_j) \tag{3.4.6}$$

under the condition that the other variables $x_r \in X_r$ $(1 \leq r \leq k,\ r \neq i,\ r \neq j)$ are assumed to be fixed.

Lemma 3.4.3: *Let $H : X_1 \times \ldots \times X_k \to GL_n(\mathbb{K})$, where $X_1,\ldots, X_k$ are $k \geq 3$ nonempty sets. The mapping H has a factorization* (3d) *if and only if each partial mapping H_{ij} $(1 \leq i < j \leq k$,* see (3.4.6)) *is of the form*

$$H_{ij}(x_i, x_j) = \Phi(x_i) \cdot \Psi(x_j) \quad (x_i \in X_i,\ x_j \in X_j)\,, \tag{3.4.7}$$

for each $(k-2)$-tuple of the other variables $x_r \in X_r$ $(1 \leq r \leq k,\ r \neq i,\ r \neq j)$.

Proof: If H is as in (3d) and $1 \leq i < j \leq k$, then (3.4.7) obviously holds with

$$\Phi(x_i) = F_1(x_1) \cdot \ldots \cdot F_i(x_i) \quad \text{and} \quad \Psi(x_j) = F_{i+1}(x_{i+1}) \cdot \ldots \cdot F_k(x_k)\,.$$

Conversely, suppose that each $\binom{k}{2}$-tuple of partial functions H_{ij} of a given function H satisfies (3.4.7). Choose some elements $u_r \in X_r$, $1 \leq r \leq k-1$. In view of the rule (2.4.5) applied to each H_{ij}, the relations

$$F_i(x_i) := H(u_1, \ldots, u_{i-1}, x_i, \ldots, x_k) \cdot H^{-1}(u_1, \ldots, u_i, x_{i+1}, \ldots, x_k) \qquad (1 \leq i \leq k-1)$$

determine $k-1$ mappings $F_i : X_i \to GL_n(\mathbb{K})$ (in a correct way). Moreover, the identity $F_1(x_1) \cdot \ldots \cdot F_i(x_i) = H(x_1, \ldots, x_k) \cdot H^{-1}(u_1, \ldots, u_i, x_{i+1}, \ldots, x_k)$ holds for $i = 1, 2, \ldots, k-1$. So putting $F_k(x_k) = H(u_1, \ldots, u_{k-1}, x_k)$, we get some factorization (3d). □

Theorem 3.4.4 [Ši 4]: *Let $H : I_1 \times \ldots \times I_k \to GL_n(\mathbb{K})$ be a differentiable mapping, where $I_1, \ldots, I_k$ are intervals in $\mathbb{R}$. Then H has a factorization* (3d) *on $I_1 \times \ldots \times I_k$ if and only if H solves the system of $\binom{k}{2}$ differential equations*

$$H_{x_i x_j} = H_{x_i} \cdot H^{-1} \cdot H_{x_j} \quad (1 \leq i < j \leq k) \quad \textit{on } I_1 \times \ldots \times I_k\,.$$

Proof: In view of Lemma 3.4.3, the proof follows immediately from Theorem 2.4.5, appplied to each factorization (3.4.7). □

Remarks 3.4.5: **(i)** The reader may ask whether the system of $\binom{k}{2}$ partial factorizations (3.4.7) in the statement of Lemma 3.4.3 can be reduced to a subsystem, say the subsystem of $(k-1)$ factorizations (3.4.7), with

$$(i, j) \in \{(1,2), (2,3), \ldots, (k-1, k)\}\,.$$

The negative answer follows from the following example. Given indices p and q $(1 \leq p < q \leq k)$, define a mapping $H = H(x_1, \ldots, x_k) = \Phi(x_p, x_q)$, where Φ is a matrix-valued function of two variables which does not permit any factorization

(2a). It is obvious that such a mapping H has $\binom{k}{2}$ factorizations (3.4.7) with only one exception, namely for $i=p$ and $j=q$.

(ii) Let us pose an *open problem* which generalizes the factorization problem (3d): *Given k surjective mappings $\varphi_i : X \to Y_i$, $1 \le i \le k$, find some necessary and sufficient conditions in order for a mapping $H : X \to GL_n(\mathbb{K})$ to be factorizable into*

$$H(x) = F_1(\varphi_1(x)) \cdot F_2(\varphi_2(x)) \cdot \ldots \cdot F_k(\varphi_k(x)) \tag{3.4.8}$$

with some factors $F_i : Y_i \to GL_n(\mathbb{K})$. We are able to solve the problem only in the case when the mapping $\varphi : X \to Y_1 \times Y_2 \times \ldots \times Y_k$ defined by

$$\varphi(x) = (\varphi_1(x), \varphi_2(x), \ldots, \varphi_k(x)) \quad (x \in X)$$

is a bijection. Then (3.4.8) can be reduced by the transformation $\tilde{H}(y) = H(\varphi^{-1}(y))$ to a problem which is treated here: $\tilde{H}(y_1, y_2, \ldots, y_k) = F_1(y_1) \cdot F_2(y_2) \cdot \ldots \cdot F_k(y_k)$. (An example of this procedure was given in Corollary 2.4.7.)

Now we turn our attention to the problem involving a smooth nonsingular matrix function H of $p+q$ real variables which permits a factorization

$$H(x_1, \ldots, x_p; y_1, \ldots, y_q) = F(x_1, \ldots, x_p) \cdot G(y_1, \ldots, y_q)\,. \tag{3.4.9}$$

Let us emphasize the fact that if $H : (X_1 \times \ldots \times X_p) \times (Y_1 \times \ldots \times Y_q) \to GL_n(\mathbb{K})$, where X_i and Y_j are any nonempty sets, then Theorem 2.4.1 with the variables $\boldsymbol{x} = (x_1, \ldots, x_p)$ and $\boldsymbol{y} = (y_1, \ldots, y_q)$ yields the following conclusion: *the factors F and G from any factorization* (3.4.9) *of the function H are given by*

$$\begin{aligned} F(x_1, \ldots, x_p) &= H(x_1, \ldots, x_p; v_1, \ldots, v_q) \cdot C \\ G(y_1, \ldots, y_q) &= D \cdot H(u_1, \ldots, u_p; y_1, \ldots, y_q) \end{aligned} \tag{3.4.10}$$

where the elements $u_i \in X_i$ and $v_j \in Y_j$ are chosen arbitrarily and the matrices $C, D \in GL_n(\mathbb{K})$ satisfy $C \cdot D = H^{-1}(u_1, \ldots, u_p; v_1, \ldots, v_q)$.

Theorem 3.4.6: *Let $X = X_1 \times \ldots \times X_p$ and $Y = Y_1 \times \ldots \times Y_q$ be the Cartesian products of real intervals $X_1, \ldots, X_p$ and $Y_1, \ldots, Y_q$, respectively. Suppose that a mapping $H : X \times Y \to GL_n(\mathbb{K})$ has the partial derivatives H_{x_i}, H_{y_j} and $H_{x_i y_j}$ (in some order of differentiation) on the set $X \times Y$, $1 \le i \le p$ and $1 \le j \le q$. Then the mapping H has a factorization* (3.4.9) *if and only if H satisfies the system of pq differential equations*

$$H_{x_i y_j} = H_{x_i} \cdot H^{-1} \cdot H_{y_j} \quad (1 \le i \le p,\ 1 \le j \le q) \quad \textit{on the set } X \times Y. \tag{3.4.11}$$

Proof: Consider the partial mappings $H_{ij} : X_i \times Y_j \to GL_n(\mathbb{K})$ defined by

$$H_{ij}(x_i, y_j) := H(x_1, \ldots, x_p; y_1, \ldots, y_q)$$

subject to the condition that the other variables

$$x_1, \ldots, x_{i-1}, x_{i+1}, \ldots, x_p \quad \text{and} \quad y_1, \ldots, y_{j-1}, y_{j+1}, \ldots, y_q$$

are assumed to be fixed.

(i) If H is as in (3.4.9), then $H_{ij}(x_i, y_j) = F_i(x_i) \cdot G_j(y_j)$, where

$$F_i(x_i) = F(x_1, \ldots, x_p) \quad \text{and} \quad G_j(y_j) = G(y_1, \ldots, y_q) \, .$$

Applying Theorem 2.4.5 (or Remark 2.4.6) to each function H_{ij}, we conclude that H satisfies (3.4.11).

(ii) Suppose that H solves (3.4.11). Then Theorem 2.4.5 (or Remark 2.4.6) implies that each partial function H_{ij} is of type (2b) on the set $X_i \times Y_j$. Choose $u_1 \in X_1$ and $\boldsymbol{v} \in Y$ and define a mapping $\Phi_1 : X \to GL_n(\mathbb{K})$ by

$$\Phi_1(\boldsymbol{x}) := H(x_1, \ldots, x_p; \boldsymbol{v}) \cdot H^{-1}(u_1, x_2, \ldots, x_p; \boldsymbol{v})$$

for each $\boldsymbol{x} = (x_1, \ldots, x_p) \in X$. According to the rule (2.4.5) applied to H_{1j}, where $1 \leq j \leq q$, the matrix product

$$H(x_1, \ldots, x_p; y_1, \ldots, y_q) \cdot H^{-1}(u_1, x_2, \ldots, x_p; y_1, \ldots, y_q)$$

depends on none of the variables $y_1, \ldots, y_q$, i.e. it equals to $\Phi(x_1, \ldots, x_p)$. This leads to the factorization

$$H(\boldsymbol{x}; \boldsymbol{y}) = \Phi_1(\boldsymbol{x}) \cdot H(u_1, x_2, \ldots, x_p; \boldsymbol{y})$$

for each $\boldsymbol{x} = (x_1, \ldots, x_p) \in X$ and $\boldsymbol{y} \in Y$. In the case when $p > 1$, we repeat the previous procedure to the function $\tilde{H}(x_2, \ldots, x_p; \boldsymbol{y}) = H(u_1, x_2, \ldots, x_p; \boldsymbol{y})$ to obtain the factorization

$$H(u_1, x_2, \ldots, x_p; \boldsymbol{y}) = \Phi_2(x_2, \ldots, x_p) \cdot H(u_1, u_2, x_3, \ldots, x_p; \boldsymbol{y})$$

(with a chosen $u_2 \in X_2$), etc. After p repetitions we conclude that H is of the form (3.4.9), in which

$$F(x_1, \ldots, x_p) = \Phi_1(x_1, \ldots, x_p) \cdot \Phi_2(x_2, \ldots, x_p) \cdot \ldots \cdot \Phi_p(x_p)$$

and $G(y_1, \dots, y_q) = H(u_1, \dots, u_p; y_1, \dots, y_q)$. This completes the proof. □

Let us finish this section by an application of the preceding factorization theorem to the theory of webs[4] (cf. [Gob]). We will present here a new approach to the classical problem of finding conditions under which a 3-web on a smooth $2n$-dimensional manifold is locally equivalent with a web formed by three systems of parallel n-planes in $\mathbb{R}^{2n}$.

Local properties of multicodimensional differentiable 3-webs are usually studied by methods of adapted co-frames, Pfaffian equations, and Cartan methods, or by the apparatus of coordinatizing local differentiable quasigroups. In most works (especially by Russian authors), the so-called closure conditions for webs are formulated as requirements on curvature and torsion tensors of an appropriate connection. We apply here a dual viewpoint working with tangent distributions to the leaves of foliations defining a 3-web, and with corresponding projector fields.

An ordered differentiable 3-web $\mathscr{W}$ on a C^∞ (smooth) manifold M_{2n} of dimension $2n$ can be regarded as a triple $\mathscr{W} = (D_1, D_2, D_3)$ of (smooth) n-dimensional integrable distributions which are pairwise complementary. The tangent bundle T M can be decomposed as a Whitney sum:

$$\mathrm{T}M = D_\alpha \oplus D_\beta, \qquad (\alpha \neq \beta;\ \alpha, \beta \in \{1, 2, 3\}).$$

These decompositions determine web projectors P_α^β associated with $\mathscr{W}$, and we introduce $(1,1)$-tensor fields $B_\gamma = P_\gamma^\alpha - P_\beta^\gamma$ satisfying $B_\gamma^2 = \mathrm{id}$, where id is the identity automorphism. The pairs $[D_\alpha, D_\beta]$ are integrable almost product structures.

As mentioned above, we are interested in conditions under which a 3-web is locally diffeomorphic with a web formed by three systems of parallel n-planes in $\mathbb{R}^{2n}$. Such webs, which satisfy the so-called Thomsen closure condition, can be coordinatized by commutative Lie groups, and are usually called *parallelizable*. We want to substitute a Chern connection (and its curvature and torsion tensor) which is more or less an additional object, by projectors (and tensor fields constructed from web projectors) which are naturally related to the web, or can be even regarded as 1-forms defining a 3-web.

Our computations will use local coordinates *adapted* to the almost product structure $[D_1, D_2]$, i.e. coordinates $(U; x_1, \dots, x_{2n})$ on a neighbourhood U for which the base vector fields $\partial_i = \partial/\partial x_i$ $(i = 1, \dots, n)$ span D_1 on U, and $\partial_{n+i} = \partial/\partial x_{n+i}$ span D_2 on U. In an adapted chart, the endomorphisms $(P_3^1)_x$, $(B_3)_x$ of the tangent space $\mathrm{T}_x M$ have block-matrix representation

$$P_1^3 = \begin{pmatrix} E_n & 0 \\ -P & 0 \end{pmatrix} \quad \text{and} \quad B_3 = \begin{pmatrix} 0 & P^{-1} \\ P & 0 \end{pmatrix}. \tag{3.4.12}$$

[4]We wish to thank Dr A. Vanžurová for providing this application.

Here P is a nonsingular $n \times n$ matrix depending on $x_1, \ldots, x_{2n}$, while E_n and 0 denote the unit and zero $n \times n$ matrices, respectively. We can say that a 3-web is *parallelizable* if there exist local $[D_1, D_2]$-adapted coordinates (i.e. ∂_i span D_1, ∂_{n+i} span D_2) such that $\partial_i + \partial_{n+i}$ span D_3, $i = 1, \ldots, n$. Since the almost product structure $[D_1, D_2]$ is integrable, there always exists a coordinate transformation f (via Frobenius' theorem) which transforms adapted coordinates $(x_i; x_{n+i})$ into coordinates $(x'_i; x'_{n+i})$ which are again adapted (that is, both the distributions D_1 and D_2 are fixed under f):

$$\begin{aligned} x'_i &= f^i(x_1, \ldots, x_n) \\ x'_{n+i} &= f^{n+i}(x_{n+1}, \ldots, x_{2n}) \end{aligned} \qquad (1 \le i \le n). \tag{3.4.13}$$

This means that

$$\frac{\partial f^i}{\partial x_{n+j}} = 0 \quad \text{and} \quad \frac{\partial f^{n+i}}{\partial x_j} = 0 \quad (1 \le i \le n,\ 1 \le j \le n).$$

Theorem 3.4.7 [Va]: *A 3-web is parallelizable if and only if the matrix P from representation* (3.4.12) *satisfies conditions*

$$(P_{x_{n+j}} P^{-1})_{x_i} = 0 \tag{3.4.14}$$

$$(P_{x_{n+i}})^k_j = (P_{x_{n+j}})^k_i \quad \textit{and} \quad (P^{-1}_{x_i})^k_j = (P^{-1}_{x_j})^k_i \tag{3.4.15}$$

for each $i, j, k \in \{1, 2, \ldots, n\}$. Here $P^{-1}_{x_i} = \frac{\partial}{\partial x_i}(P^{-1})$ and $(M)^k_i$ denotes the entry of the matrix M lying in the i-th row and k-th column.

Proof: Given transformation (3.4.13), define the Jacobi matrices

$$F = \left[\frac{\partial f^i}{\partial x_j}\right]^n_{i,j=1} \quad \text{and} \quad \tilde{F} = \left[\frac{\partial f^{n+i}}{\partial x_{n+j}}\right]^n_{i,j=1}.$$

Similarly, for the inverse transformation

$$\begin{aligned} x_i &= g^i(x'_1, \ldots, x'_n) \\ x_{n+i} &= g^{n+i}(x'_{n+1}, \ldots, x'_{2n}) \end{aligned} \qquad (1 \le i \le n)$$

define

$$G = \left[\frac{\partial g^i}{\partial x'_j}\right]^n_{i,j=1} \quad \text{and} \quad \tilde{G} = \left[\frac{\partial g^{n+i}}{\partial x'_{n+j}}\right]^n_{i,j=1}.$$

The tangent vectors $\dfrac{\partial}{\partial x'_i} + \dfrac{\partial}{\partial x'_{n+i}}$ span D_3 if and only if

$$P_1^3\left(\frac{\partial}{\partial x'_i} + \frac{\partial}{\partial x'_{n+i}}\right) = 0 \qquad (i = 1, 2, \ldots, n).$$

An easy computation shows that this requirement is equivalent to

$$\frac{\partial g^{n+s}}{\partial x'_{n+j}}(P)_s^k - \frac{\partial g^k}{\partial x'_j} = 0 \qquad (i, j, k \in \{1, 2, \ldots, n\}),$$

which can be rewritten in the following matrix form

$$\tilde{G} \cdot P = G\,, \quad \text{i.e.} \quad P = \tilde{G}^{-1} \cdot G\,.$$

Since $\tilde{G}^{-1} = \tilde{F}$ and $G = F^{-1}$ in the corresponding coordinates, we therefore obtain the factorization

$$P(x_1, x_2, \ldots, x_{2n}) = A(x_{n+1}, \ldots, x_{2n}) \cdot B(x_1, \ldots, x_n) \tag{3.3.16}$$

with $A = \tilde{F}$ and $B = F^{-1}$. In view of Theorem 3.4.6, the matrix P admits some factorization of type (3.3.16) if and only if

$$P_{x_{n+j}x_i} = P_{x_{n+j}} \cdot P^{-1} \cdot P_{x_i} \quad (i, j \in \{1, 2, \ldots, n\})\,,$$

which is equivalent to (3.4.14). Moreover, by formula (3.4.10), the factors A and B from any factorization (3.4.16) are given by

$$\begin{aligned} A(x_{n+1}, \ldots, x_{2n}) &= P(u_1, \ldots, u_n, x_{n+1}, \ldots, x_{2n}) \cdot C \\ B(x_1, \ldots, x_n) &= D \cdot P(x_1, \ldots, x_n, u_{n+1}, \ldots, u_{2n}) \end{aligned}$$

where $u = (u_1, \ldots, u_{2n})$ is a constant point (arbitrarily chosen in the coordinate neighbourhood U) and the nonsingular $n \times n$ matrices C and D satisfy $C \cdot D = P^{-1}(u_1, \ldots, u_{2n})$. It is well known that the entries of the matrices A and B can be written as partial derivatives

$$(A)_j^i = \frac{\partial f^{n+i}}{\partial x_{n+j}}, \quad (B)_j^i = \frac{\partial g^i}{\partial x'_j}, \quad \text{or} \quad (B^{-1})_j^i = \frac{\partial f^i}{\partial x_j}$$

if and only if

$$\frac{\partial (A)_i^k}{\partial x_{n+j}} = \frac{\partial (A)_j^k}{\partial x_{n+i}} \qquad \text{and} \qquad \frac{\partial (B^{-1})_i^k}{\partial x_j} = \frac{\partial (B^{-1})_j^k}{\partial x_i}$$

for each $i, j, k \in \{1, 2, \ldots, n\}$. These n^3 conditions can be rewritten as

$$\begin{cases} \sum_{s=1}^{n}((P_{x_{n+j}})_i^s - (P_{x_{n+i}})_j^s)(C)_s^k = 0 \\ \sum_{s=1}^{n}((P_{x_{n+j}}^{-1})_i^s - (P_{x_{n+i}}^{-1})_j^s)(D^{-1})_s^k = 0 \end{cases}$$

which is a system equivalent to (3.4.15), because the matrices C and D^{-1} are nonsingular. □

Remarks 3.4.8: (i) Parallelizability of a 3-web means, in fact, simultaneous integrability of almost product structures $[D_\alpha, D_\beta]$, or simultaneous integrability of polynomial structures P and B satisfying the following polynomial equations

$$P^2 - P = 0, \qquad B^2 - I = 0 \qquad \text{and} \qquad PB + BP = B$$

on a manifold.

(ii) It can be verified that the conditions (3.4.14) and (3.4.15) are equivalent to the vanishing of curvature and torsion tensors of the canonical Chern connection of a 3-web.

4 FINITE-DIMENSIONAL SPACES OF SMOOTH FUNCTIONS

The goal of this chapter is to develop a method for finding a system of linear partial differential equations whose solution set forms a prescribed finite-dimensional linear space of smooth functions in several independent variables. The results of this procedure will be used in Chapter 5 for solving the general decomposition problem (2a).

Let us start our considerations by remarking that the above-mentioned problem of describing a functional space by a system of differential equations cannot be solved "separately" in particular by the method of independent variables. For example, let us consider the three-dimensional space of functions $z\colon \mathbb{R}^2 \to \mathbb{R}$ defined by

$$S = \{z \mid z(x,y) = \gamma_1 x + \gamma_2 y + \gamma_3 xy : \gamma_1, \gamma_2, \gamma_3 \in \mathbb{R}\}\,. \tag{4a}$$

It is obvious that differential equations for functions $z \in S$ in each of the variables x and y are of the form

$$z_{xx} = 0 \quad \text{and} \quad z_{yy} = 0\,. \tag{4b}$$

However, the solution space of the system (4b)

$$S^* = \{z \mid z(x,y) = \gamma_1 x + \gamma_2 y + \gamma_3 xy + \gamma_4 : \gamma_1, \gamma_2, \gamma_3, \gamma_4 \in \mathbb{R}\}$$

forms a wider (fourth-dimensional) space. To describe the original space S, we have to append to (4b) another differential equation. Using the general theory developed in Section 4.2, we will show that the third required equation is of the form

$$xy \cdot z_{xy} - x \cdot z_x - y \cdot z_y + z = 0$$

(see part (i) of Examples 4.2.1).

4.1. General Wronski matrices

Throughout Chapter 4, we will consider scalar-valued functions in k independent real variables $x_1, x_2, \dots, x_k$, defined in some *region* $X \subseteq \mathbb{R}^k$ (i.e. nonempty, open and connected subset of $\mathbb{R}^k$). For the sake of simplicity, all functions under consideration are assumed to be of class $C^\infty(X, \mathbb{K})$. Each partial derivative

$$D = \frac{\partial^{\alpha_1+\alpha_2+\cdots+\alpha_k}}{\partial x_1^{\alpha_1} \partial x_2^{\alpha_2} \dots \partial x_k^{\alpha_k}}, \tag{4.1.1}$$

where $\alpha_1, \dots, \alpha_k$ are any nonnegative integers, will be considered to be an operator $D\colon C^\infty(X, \mathbb{K}) \to C^\infty(X, \mathbb{K})$; the set of all these operators D will be denoted by $\mathscr{D}(X)$. (In the case when $\alpha_1 = \cdots = \alpha_k = 0$, the operator (4.1.1) becomes the identity $D = \mathrm{id}$.) By $\mathscr{D}^{(1)}(X)$ we denote the k-element subset of $\mathscr{D}(X)$ consisting of all the partial derivatives of order 1:

$$\mathscr{D}^{(1)}(X) := \left\{ \frac{\partial}{\partial x_1}, \frac{\partial}{\partial x_2}, \dots, \frac{\partial}{\partial x_k} \right\}.$$

For each m-tuple of derivatives $D_1, D_2, \dots, D_m \in \mathscr{D}(X)$ and for each n-tuple of functions $f_1, f_2, \dots, f_n \in C^\infty(X, \mathbb{K})$, the following matrix

$$W_{m,n}[D_1, \dots, D_m; f_1, \dots, f_n] := \begin{pmatrix} D_1 f_1 & D_1 f_2 & \dots & D_1 f_n \\ D_2 f_1 & D_2 f_2 & \dots & D_2 f_n \\ \vdots & \vdots & \ddots & \vdots \\ D_m f_1 & D_m f_2 & \dots & D_m f_n \end{pmatrix} \tag{4.1.2}$$

is called the *Wronski matrix of size* $m \times n$. For the sake of simplicity, in the case $m = n$, we will denote the (square) matrix (4.1.2) by $W_n[D_i; f_j]$. Finally, instead of

$$W_{n+1,n+1}[D_1, \dots, D_n, D; f_1, \dots, f_n, f]$$

we will write $W_{n+1}[D_i, D; f_j, f]$.

Now we are ready to state and prove the basic result of this chapter, which generalizes Theorem 1.2.1. For some versions of this result, see [O], [Kr], [W 2].

Theorem 4.1.1 [ČŠ 2]: *Let $f_1, f_2, \dots, f_n, f \in C^\infty(X, \mathbb{K})$, where X is a region in $\mathbb{R}^k$.*

(i) *If the functions $f_1, f_2, \dots, f_n$ are linearly dependent in X, then*

$$\det W_n[D_i; f_j](x) = 0 \quad (\text{at each point } x \in X) \tag{4.1.3}$$

for any n-tuple of derivatives $D_1, D_2, \dots, D_n \in \mathscr{D}(X)$.

(ii) *If there exist derivatives* $D_1 = \mathrm{id}, D_2, \dots, D_n \in \mathscr{D}(X)$ *such that*

$$\det W_n[D_i; f_j](x) \neq 0 \quad (\textit{at each point } x \in X) \tag{4.1.4}$$

and that, for each $D \in \mathscr{D}^{(1)}(X)$ *and each* $r = 1, 2, \dots, n$,

$$\det W_{n+1}[D_i, D \circ D_r; f_j, f](x) = 0 \quad (\textit{at each point } x \in X) \tag{4.1.5}$$

then the function f *is a linear combination of* $f_1, \dots, f_n$ *in the region* X.

Proof: (i) If there exists a nonzero vector $\vec{c} = (\gamma_1, \dots, \gamma_n) \in \mathbb{K}^n$ such that

$$\gamma_1 f_1 + \gamma_2 f_2 + \cdots + \gamma_n f_n = 0 \quad (\text{on } X)$$

then the identity

$$\gamma_1 D f_1 + \gamma_2 D f_2 + \cdots + \gamma_n D f_n = 0 \quad (\text{on } X)$$

holds for each $D \in \mathscr{D}(X)$ as well. Thus $\vec{c}$ is a nonzero vector orthogonal to each row of the matrix $W_n[D_i; f_j]$ and hence this matrix is singular, which proves (4.1.3).

(ii) For each fixed $r = 1, 2, \dots, n$, the conditions (4.1.4) and (4.1.5) imply that the last column of $W_{n+1}[D_i, D \circ D_r; f_j, f]$ is a linear combination of the previous ones:

$$D_i f(x) = \sum_{s=1}^{n} b_s(x) \cdot D_i f_s(x) \quad (1 \le i \le n,\ x \in X) \tag{4.1.6}$$

and

$$D \circ D_r f(x) = \sum_{s=1}^{n} b_s(x) \cdot D \circ D_r f_s(x) \quad (x \in X). \tag{4.1.7}$$

The coefficients b_s can be evaluated from (4.1.6) by means of Cramer's rule:

$$b_s(x) = \frac{\det W_n[D_i; f_1, \dots, f_{s-1}, f, f_{s+1}, \dots, f_n](x)}{\det W_n[D_i; f_j](x)} \quad (1 \le s \le n,\ x \in X).$$

This leads to the conclusion that the functions b_s belong to $C^\infty(X, \mathbb{K})$ and they do not depend on the index $r = 1, 2, \dots, n$. Applying now any operator $D \in \mathscr{D}^1(X)$ to both sides of (4.1.6) and then subtracting (4.1.7) with $r = i$, we obtain

$$0 = \sum_{s=1}^{n} D b_s(x) D_i f_s(x) \quad (1 \le i \le n,\ x \in X).$$

In view of (4.1.4), the last equality means that $D b_s = 0$ on X, for each $D \in \mathscr{D}^1(X)$ and each $s = 1, \dots, n$; hence all the coefficients b_s are constant on the set X. Since, in addition, $D_1 = \mathrm{id}$, equality (4.1.6) with $i = 1$ implies that the function f is a linear combination of $f_1, \dots, f_n$ in the set X. $\square$

4.2. Special systems of PDE's

Part (ii) of Theorem 4.1.1 provides a background for a procedure of finding systems of differential equations with a prescribed finite-dimensional solution space

$$S = \{f \mid f = \gamma_1 f_1 + \gamma_2 f_2 + \dots + \gamma_n f_n : \gamma_1, \gamma_2, \dots, \gamma_n \in \mathbb{K}\}.$$

Assuming the basic functions $f_1, f_2, \dots, f_n \in C^\infty(X, \mathbb{K})$ to satisfy (4.1.4) with a suitable n-tuple of derivatives $D_1 = \mathrm{id}, D_2, \dots, D_n \in \mathscr{D}(X)$, Theorem 4.1.1 implies that the space S consists of exactly those functions $f \in C^\infty(X, \mathbb{K})$ that satisfy the system of nk conditions

$$\det W_{n+1}[D_i, D \circ D_r; f_j, f](x) = 0 \quad \text{(at each point } x \in X) \tag{4.2.1}$$

in which

$$D \in \mathscr{D}^{(1)}(X) = \left\{ \frac{\partial}{\partial x_1}, \frac{\partial}{\partial x_2}, \dots, \frac{\partial}{\partial x_k} \right\} \quad \text{and} \quad r = 1, 2, \dots, n.$$

Let us expand the determinant in the left-hand side of (4.2.1) with respect to the last column, formed by elements $D_1 f, D_2 f, \dots, D_n f, D \circ D_r f$. In this expansion, the element $D \circ D_r f$ is multiplied by a nonzero coefficient, because of (4.1.4). Therefore, we may rewrite the system (4.2.1) into the form

$$\left(\frac{\partial}{\partial x_i} \circ D_r \right) f = \sum_{s=1}^{n} a_{irs}(x) \cdot D_s f \quad (i = 1, 2, \dots, k;\ r = 1, 2, \dots, n) \tag{4.2.2}$$

in which the coefficients a_{irs} are uniquely determined by

$$a_{irs} = \frac{\det W_n[D_1, \dots, D_{s-1}, \frac{\partial}{\partial x_i} \circ D_r, D_{s+1}, \dots, D_n; f_j]}{\det W_n[D_i; f_j]}. \tag{4.2.3}$$

In fact, formulas (4.2.3) follow from Cramer's rule applied to the system of n equations (4.2.2) with $f = f_1, f_2, \dots, f_n$. It is easily seen from (4.2.3) that all the coefficients a_{irs} are of class $C^\infty(X, \mathbb{K})$ and their values *do not depend* on the choice of the basis $f_1, f_2, \dots, f_n$ of the given functional space S. (If we change the basis, then both numerator and denominator of the fraction in the right-hand side of (4.2.3) are multiplied by the same constant). Consequently, the system of differential equations (4.2.2) with the given solution space S depends only on the choice of the "basic" n-tuple of derivatives $D_1 = \mathrm{id}, D_2, \dots, D_n \in \mathscr{D}(X)$; this choice will be disscussed in Section 4.3. Throughout the chapter, each system (4.2.2) will, for brevity, be called a *PDE system*. Let us emphasize that some of the nk equations in a PDE system (4.2.2) may be trivial: if r is an index such that the operator $\frac{\partial}{\partial x_i} \circ D_r$ belongs to the n-tuple $D_1, D_2, \dots, D_n$, then (4.2.3) implies that

$$a_{irs} = \begin{cases} 1 & \text{if } \frac{\partial}{\partial x_i} \circ D_r = D_s \\ 0 & \text{for the other values of } s. \end{cases}$$

Thus in the case $k = 1$, when $\mathscr{D}^{(1)}(X) = \{\frac{d}{dx}\}$ and when we usually set $D_i = \frac{d^{i-1}}{dx^{i-1}}$ for each $i = 1, \dots, n$, there exists only one nontrivial equation in (4.2.2):

$$f^{(n)} = \sum_{s=0}^{n-1} a_s(x) f^{(s)} . \tag{4.2.4}$$

Examples 4.2.1: **(i)** Let us start with the example

$$S = \{z \mid z(x,y) = \gamma_1 x + \gamma_2 y + \gamma_3 xy : \gamma_1, \gamma_2, \gamma_3 \in \mathbb{R}\}$$

from the introductory part of this chapter. The triple of functions $z_1(x,y) = x$, $z_2(x,y) = y$ and $z_3(x,y) = xy$ forms a basis of S. Set $D_1 = \mathrm{id}$, $D_2 = \frac{\partial}{\partial x}$ and $D_3 = \frac{\partial}{\partial y}$, then the Wronskian satisfies

$$\det W_3[D_i; z_j](x,y) = \begin{vmatrix} x & y & xy \\ 1 & 0 & y \\ 0 & 1 & x \end{vmatrix} = -xy .$$

Thus the condition $\det W_3[D_i; z_j] \neq 0$ is fulfilled, for example, in the first quadrant $X = \{(x,y) : x > 0, y > 0\}$. Consequently, the space S on X can be described by the system (4.2.2). Although this system formally consists of $3 \times 2 = 6$ equations whose left-hand sides are sucessively

$$\frac{\partial}{\partial x}, \frac{\partial^2}{\partial x \partial y}, \frac{\partial^2}{\partial x^2}, \frac{\partial}{\partial y}, \frac{\partial^2}{\partial y \partial x}, \frac{\partial^2}{\partial y^2},$$

exactly three (distinct) of them are nontrivial and are of the form

$$z_{xx} = 0, \quad z_{yy} = 0 \quad \text{and} \quad z_{xy} = \frac{z_x}{y} + \frac{z_y}{x} - \frac{z}{xy}.$$

To derive these equations, it suffices to equate to zero each of the determinants

$$\begin{vmatrix} x & y & xy & z \\ 1 & 0 & y & z_x \\ 0 & 1 & x & z_y \\ 0 & 0 & 0 & z_{xx} \end{vmatrix}, \quad \begin{vmatrix} x & y & xy & z \\ 1 & 0 & y & z_x \\ 0 & 1 & x & z_y \\ 0 & 0 & 0 & z_{yy} \end{vmatrix} \quad \text{and} \quad \begin{vmatrix} x & y & xy & z \\ 1 & 0 & y & z_x \\ 0 & 1 & x & z_y \\ 0 & 0 & 1 & z_{xy} \end{vmatrix}.$$

It is clear that the same system describes the space S on the second, third and fourth quadrants, respectively. Notice also that the coordinate axes are formed by the *singular points* of the third equation (for the mixed derivative z_{xy}).

(ii) The space of functions

$$S = \{z \mid z(x,y) = \gamma_1 + \gamma_2 xy^2 + \gamma_3 x^2 y : \gamma_1, \gamma_2, \gamma_3 \in \mathbb{R}\}$$

can be described by the system of equations

$$\begin{cases} z_{xx} = \dfrac{4}{3x} z_x - \dfrac{2y}{3x^2} z_y \\ z_{xy} = \dfrac{2}{3y} z_x + \dfrac{2}{3x} z_y \\ z_{yy} = \dfrac{-2x}{3y^2} z_x + \dfrac{4}{3y} z_y \end{cases}$$

in each of the four coordinate quadrants. We leave this fact for the reader to verify.

(iii) Now we introduce an example showing that the choice of the suitable "basic" derivatives D_i for system (4.2.2) may differ in different subregions of the domain of definition and that the subsystem of nontrivial equations (4.2.2) can be reduced again. For this purpose, let us consider an n-dimensional functional space

$$S_n = \{z \mid z(x,y) = \gamma_1 + \gamma_2 xy + \cdots + \gamma_n x^{n-1} y^{n-1}\}$$

with the basis $z_j(x,y) = (xy)^{j-1}$, $j = 1, \ldots, n$. For $D_i = \frac{\partial^{i-1}}{\partial x^{i-1}}$ we have

$$\det W_n[D_i; z_j] = \begin{vmatrix} 1 & xy & x^2y^2 & \ldots & x^{n-1}y^{n-1} \\ 0 & y & 2y^2x & \ldots & (n-1)x^{n-2}y^{n-2} \\ \vdots & \vdots & \vdots & \ddots & \vdots \\ 0 & 0 & 0 & \ldots & (n-1)!y^{n-1} \end{vmatrix}$$
$$= 1!2!\ldots(n-1)!y^{\frac{(n-1)n}{2}} \neq 0 \text{ in the "upper" half-plane } X = \{(x,y) | y > 0\}.$$

Thus the space S_n on X can be described by the system (4.2.2) in which exactly

$n+1$ distinct equations are nontrivial and are of the form

$$\begin{cases} \dfrac{\partial^n z}{\partial x^n} = 0 \\ \dfrac{\partial^{r+1} z}{\partial x^r \partial y} = \displaystyle\sum_{s=0}^{n-1} a_{rs}(x,y) \frac{\partial^s z}{\partial x^s} \quad (r=0,1,\ldots,n-1)\,. \end{cases} \tag{4.2.5}$$

To compute the coefficients a_{rs}, we adapt a different procedure for the direct calculations of determinants in (4.2.3). Note that for each $z \in S_n$, the following identity holds:

$$\frac{\partial z}{\partial y} = \frac{\partial}{\partial y}\left(\sum_{j=0}^{n-1} \gamma_{j+1}(xy)^j\right) = \sum_{j=0}^{n-1} j\gamma_{j+1}x^j y^{j-1} = \frac{x}{y} \cdot \frac{\partial z}{\partial x}\,. \tag{4.2.6}$$

Differentiating (4.2.6), we obtain the other equations in (4.2.5):

$$\frac{\partial^{r+1} z}{\partial x^r \partial y} = \frac{\partial}{\partial x^r}\left(\frac{x}{y} \cdot \frac{\partial z}{\partial x}\right) = \frac{x}{y} \cdot \frac{\partial^{r+1} z}{\partial x^{r+1}} + \frac{r}{y} \cdot \frac{\partial^r z}{\partial x^r} \quad (r=1,2,\ldots,n-2)$$

and

$$\frac{\partial^n z}{\partial x^{n-1} \partial y} = \frac{n-1}{y} \cdot \frac{\partial^{n-1} z}{\partial x^{n-1}}\,.$$

Consequently, the system (4.2.5) is equivalent to the pair of equations

$$\frac{\partial^n z}{\partial x^n} = 0 \quad \text{and} \quad \frac{\partial z}{\partial y} = \frac{x}{y} \cdot \frac{\partial z}{\partial x}\,. \tag{4.2.7}$$

The second equation in (4.2.7) is singular for $y=0$. To describe the same space S_n in another region, say, in the "left-hand" half-plane $Y = \{(x,y)|x < 0\}$, we will have to change the choice of D_i. In view of symmetry, it is clear that we may set $D_i = \frac{\partial^{i-1}}{\partial y^{i-1}}$ and derive the system of two equations

$$\frac{\partial^n z}{\partial y^n} = 0 \quad \text{and} \quad \frac{\partial z}{\partial x} = \frac{y}{x} \cdot \frac{\partial z}{\partial y}\,.$$

Finally, it is worth remarking that for this space S_n, both first-order derivatives $\frac{\partial}{\partial x}$ and $\frac{\partial}{\partial y}$ cannot be simultaneously among the elements D_i satisfying the basic nonvanishing condition (4.1.4) on an open subset of the plane. The reason for this is the second equation in (4.2.7).

(iv) In contrast to (iii), we now construct an n-tuple $f_1, \ldots, f_n$ lying in the space $C^\infty(\mathbb{R}^k, \mathbb{R})$ and possessing a curious property: the Wronskian $\det W_n[D_i; f_j]$ has no zero value, for each n-tuple of pairwise distinct derivatives $D_1, \dots, D_n \in \mathscr{D}(\mathbb{R}^k)$. The construction of such an n-tuple will be given recurrently for the number $n \geq 1$. Let $\vec{x} = (x_1, \dots, x_k) \in \mathbb{R}^k$ be arbitrary. For $n = 1$, we may set

$$f_1(\vec{x}) = \exp(\lambda_1 x_1 + \lambda_2 x_2 + \dots \lambda_k x_k)$$

where $\lambda_1, \dots, \lambda_k \in \mathbb{R}$ are nonzero constants, because for each partial derivative D we have

$$Df_1(\vec{x}) = \frac{\partial^{\alpha_1+\alpha_2+\dots+\alpha_k} f_1}{\partial x_1^{\alpha_1} \partial x_2^{\alpha_2} \dots \partial x_k^{\alpha_k}}(\vec{x}) = \lambda_1^{\alpha_1} \dots \lambda_k^{\alpha_k} \cdot f_1(\vec{x}) \neq 0 .$$

Assume now that we have already constructed $n-1$ functions

$$f_j(\vec{x}) = \exp(\lambda_{1j} x_1 + \lambda_{2j} x_2 + \dots \lambda_{kj} x_k) \quad (j = 1, 2, \dots, n-1) \tag{4.2.8}$$

satisfying $\det W_{n-1}[D_i; f_j](\vec{x}) \neq 0$ ($\vec{x} \in \mathbb{R}^k$), for each $(n-1)$-tuple of distinct derivatives $D_1, \dots, D_{n-1} \in \mathscr{D}(\mathbb{R}^k)$. We find the new n-th function f_n in the same form (4.2.8), with suitable (unknown for the present) constants $\lambda_{1n}, \dots, \lambda_{kn}$. For each n-tuple of distinct derivatives

$$D_i = \frac{\partial^{\alpha_{i1}+\alpha_{i2}+\dots+\alpha_{ik}}}{\partial x_1^{\alpha_{i1}} \partial x_2^{\alpha_{i2}} \dots \partial x_k^{\alpha_{ik}}} \quad (i = 1, 2, \dots, n),$$

we obviously have

$$\det W_n[D_i; f_j](\vec{x}) = \left(\sum_{i=1}^{n} c_i \lambda_{1n}^{\alpha_{i1}} \lambda_{2n}^{\alpha_{i2}} \dots \lambda_{kn}^{\alpha_{ik}} \right) \prod_{j=1}^{n} f_j(\vec{x})$$

where $c_i = c_i(D_1, \dots, D_n)$ are assumed to be nonzero constants. Each equation

$$\sum_{i=1}^{n} c_i(D_1, \dots, D_n) \lambda_{1n}^{\alpha_{i1}} \lambda_{2n}^{\alpha_{i2}} \dots \lambda_{kn}^{\alpha_{ik}} = 0 \tag{4.2.9}$$

where the (distinct) derivatives $D_1, \dots, D_n$ are fixed, determines a certain subset in the k-dimensional space of variables $\lambda_{1n}, \lambda_{2n}, \dots, \lambda_{kn}$, the measure of which equals zero. Since the collection of all n-tuples $D_1, \dots, D_n$ is a countable set, there exists such a k-tuple of values $\lambda_{1n}, \lambda_{2n}, \dots, \lambda_{kn}$ for which no equation (4.2.9) is satisfied. For such values of λ_{in}, the n-tuple $f_1, \dots, f_n$ possesses the desired property and the recurrent process is complete.

4.3. Selection of derivatives for Wronskians

Let us start by repeating the result of our considerations in Section 4.2: A linear n-dimensional functional space S with a basis $f_1, \dots, f_n \in C^\infty(X, \mathbb{K})$, where X is a region in $\mathbb{R}^k$, can be described by a PDE system (4.2.2), provided that the n-tuple of derivatives

$$D_1 = \mathrm{id}, D_2, \dots, D_n \in \mathscr{D}(X) \tag{4.3.1}$$

is chosen so that (4.1.4) holds. The system (4.2.2) is then uniquely determined by the space S and by the n-tuple (4.3.1). Thus the following important question arises: How to find the suitable n-tuple (4.3.1) satisfying (4.1.4)? Before we give an answer in the form of an "optimal" Algorithm 4.3.4, we mention a few preliminary observations which motivate us to define a special type of complete n-tuples (4.3.1).

(i) In view of part (i) of Theorem 4.1.1, condition (4.1.4) can be fulfilled only in the case when the functions $f_1, \dots, f_n$ are linearly independent *in each nonempty open subset* $U \subseteq X$. In Section 1.1, we have called such n-tuples $f_1, \dots, f_n$ to be *locally linearly independent*.

(ii) Although some n-tuple $f_1, \dots, f_n$ satisfy (4.1.4) with any n-tuple of pairwise distinct derivatives $D_1, \dots, D_n$ (see part (iv) of Examples 4.2.1), it is easily seen that for a general n-tuple $f_1, \dots, f_n$ (even a locally linearly independent one), there need not exist an n-tuple (4.3.1) satisfying (4.1.4) *globally*, i.e. in the whole domain of definition X (see parts (i)–(iii) of Examples 4.2.1).

(iii) For the case $k = 1$, when X is an open interval in $\mathbb{R}$, the following conclusion has been proved in part (i) of Remarks 1.2.2: *If $f_1, \dots, f_n \in C^{n-1}(X, \mathbb{K})$ are locally linearly independent functions, then their Wronskian is nonzero almost everywhere in X.* Consequently, in this case, the problem of selection of suitable derivatives (4.3.1) is solved: We choose the same n-tuple of derivatives

$$D_1 = \mathrm{id},\ D_2 = \frac{d}{dx}, \dots,\ D_n = \frac{d^{n-1}}{dx^{n-1}}, \tag{4.3.2}$$

independently of the particular n-tuple $f_1, \dots, f_n$. Thus each n-dimensional linear space S of smooth functions on the interval X, with a locally linearly independent basis, is described by a linear ordinary differential equation (4.2.4), whose singular points (if any) form a set of measure zero .

(iv) Examples 4.1.2 show that in the case $k \geq 2$, there is no *universal* n-tuple of derivatives (4.3.1) (excepting the obvious case $n = 1$), possessing the property analogous to that of the n-tuple (4.3.2): *the inequality* $\det W_n[D_i; f_j](x) \neq 0$ *should be satisfied for each n-tuple of locally linearly independent functions* $f_1, \dots, f_n$ of class $C^\infty(X, \mathbb{K})$.

The above four facts lead to the conclusion that in the case $k \geq 2$, the solution of the problem of selection of suitable derivations (4.3.1) should be expected to be in the following "local" form.

Theorem 4.3.1 [ČŠ 2]: *Let X be a region in $\mathbb{R}^k$ and let the n-tuple of functions $f_1, \dots, f_n \in C^\infty(X, \mathbb{K})$ be locally linearly independent. Then the set X_0 of all $x \in X$ satisfying*

$$\det W_n[D_i; f_j](x) \neq 0\,, \tag{4.3.3}$$

for some n-tuple of derivatives (4.3.1), *is open and dense in X.*

Proof: The set X_0 is open, because the left-hand side of (4.3.3) is continuous in x. Let U be a (small but nonempty) open subset of X. To prove that $U \cap X_0$ is nonempty, we utilize the following construction by recursion, based on Theorem 4.1.1. Since $f_1, \dots, f_n$ are linearly independent in U, there exists a region $V_1 \subseteq U$ on which $f_1 \neq 0$. Assume now that for some p, $p \in \{1, 2, \dots, n-1\}$, the derivatives $D_1 = \mathrm{id}, D_2, \dots, D_p$ are chosen so that $\det W_p[D_i; f_j](x) \neq 0$ for each $x \in V_p$, where $V_p \subseteq U$ is a given subregion. Then we select D_{p+1} as any of the kp derivatives $\frac{\partial}{\partial x_i} \circ D_r$ ($i \in \{1, 2, \dots, k\}, r \in \{1, 2, \dots, p\}$) satisfying

$$\det W_{p+1}[D_i; f_j](x) \neq 0 \quad \text{for some } x = x_0 \in V_p. \tag{4.3.4}$$

(Should all Wronskians in (4.3.4) vanish at each point of V_p, the function f_{p+1} would be a linear combination of $f_1, \dots, f_p$ in V_p, because of Theorem 4.1.1.) By continuity the inequality in (4.3.4) holds in a region $V_{p+1} \subseteq V_p$, a small neighbourhood of the point x_0. Resulting this recurrent procedure, we find some n-tuple (4.3.1) satisfying inequality (4.3.3) at each point of a region $V_n \subseteq U$. □

The above proof provides an algorithm for selection of a suitable n-tuple (4.3.1), which is, however, not too optimal in the following sense: due to the arbitrariness in the recurrent selection of the derivative D_{p+1}, the order of the selected derivatives may become unnecessarily high. For example, in the case $n = 4$, the possible result of the algorithm can be of the form

$$D_1 = \mathrm{id},\ D_2 = \frac{\partial}{\partial x},\ D_3 = \frac{\partial^2}{\partial x^2},\ D_4 = \frac{\partial^3}{\partial x^2 \partial y}\,.$$

According to the terminology introduced in the next section, such a collection of derivatives is not *complete*, because the derivatives $\frac{\partial}{\partial y}$ and $\frac{\partial^2}{\partial x \partial y}$ which *precede* the derivative D_4 are not present in it. In what follows we will propose an algorithm of selection of complete n-tuples (4.3.1), based on an idea of the linear (total) ordering of the set $\mathscr{D}(X)$, which is only partially ordered in the usual sense of the composition of operators (see Definition 4.3.3). The resulting Algorithm 4.3.4 yields Theorem 4.3.5, which is an improved version of Theorem 4.3.1.

Definitions 4.3.3: For each of two partial derivatives

$$D = \frac{\partial^{\alpha_1 + \alpha_2 + \dots + \alpha_k}}{\partial x_1^{\alpha_1} \partial x_2^{\alpha_2} \dots \partial x_k^{\alpha_k}} \quad \text{and} \quad \hat{D} = \frac{\partial^{\beta_1 + \beta_2 + \dots + \beta_k}}{\partial x_1^{\beta_1} \partial x_2^{\beta_2} \dots \partial x_k^{\beta_k}}\,, \tag{4.3.5}$$

we will write $D \prec \hat{D}$ (read "D precedes $\hat{D}$") if the following k inequalities $\alpha_i \leq \beta_i$ ($i=1,\dots,k$) are valid, but $\alpha_i \neq \beta_i$ for some i (i.e. $D \neq \hat{D}$). Clearly $D \prec \hat{D}$ if and only if there exists a derivative $\tilde{D} \neq \mathrm{id}$ such that $\hat{D} = \tilde{D} \circ D$.

An n-tuple of derivatives $D_1, \dots, D_n \in \mathscr{D}(X)$ is said to be *complete* in the case when $D_1 = \mathrm{id}$ and when, for each $i = 2, 3, \dots, n$, the following condition holds: if some derivative $\hat{D} \in \mathscr{D}(X)$ precedes D_i, then $\hat{D}$ lies among $D_1, \dots, D_{i-1}$:

$$\text{if} \quad \hat{D} \prec D_i, \quad \text{then} \quad \hat{D} = D_j \quad \text{for some } j \in \{1, 2, \dots, i-1\}.$$

In the case $k = 1$, (4.3.2) is the unique complete n-tuple, for each $n \geq 1$. On the other side, in the case $k > 1$, there are several complete n-tuples, for each $n \geq 2$. For example, in the case $k = 2$, there are exactly three distinct complete triples

$$\{\mathrm{id}, \frac{\partial}{\partial x}, \frac{\partial}{\partial y}\}, \quad \{\mathrm{id}, \frac{\partial}{\partial x}, \frac{\partial^2}{\partial x^2}\} \quad \text{and} \quad \{\mathrm{id}, \frac{\partial}{\partial y}, \frac{\partial^2}{\partial y^2}\}$$

(as usual, we write x and y instead of x_1 and x_2, respectively). The reason for this nonuniqueness is that $(\mathscr{D}(X), \prec)$, the collection of all derivatives with ordering $\prec$, is only a *partially ordered set*. For the purpose of Algorithm 4.3.4, we need to extend this ordering to a *linear* one. Strictly speaking, we will consider such a linear ordering $<$ of the set $\mathscr{D}(X)$ for the two properties

$$\begin{cases} \text{if} \quad D \prec \hat{D}, \quad \text{then} \quad D < \hat{D} \quad (\text{for each } D, \hat{D} \in \mathscr{D}(X)), \\ \text{if} \quad D < \hat{D}, \quad \text{then} \quad D \circ \tilde{D} < \hat{D} \circ \tilde{D} \quad (\text{for each } D, \hat{D}, \tilde{D} \in \mathscr{D}(X)). \end{cases} \tag{4.3.6}$$

to be satisfied. Such orderings surely exist, the most usual being the following one: for each derivatives $D, \hat{D}$ of the form (4.3.5), we set $D < \hat{D}$ if and only if

$$\alpha_1 + \cdots + \alpha_k \leq \beta_1 + \cdots + \beta_k$$

and, in the case of equality, if there exists an index $i = 1, 2, \dots, k-1$ such that $\alpha_i > \beta_i$ and $\alpha_j = \beta_j$ for $j = 1, 2, \dots, i-1$. Other orderings can be obtained from the above by permuting the variables $x_1, \dots, x_k$. There are also other orderings satisfying (4.3.6), for example, a lexicographic one: $D < \hat{D}$ if and only if there exists an index $i \in \{1, 2, \dots, n\}$ such that $\alpha_i < \beta_i$ and (if $i > 1$) $\alpha_j = \beta_j$, $j = 1, 2, \dots, i-1$.

Algorithm 4.3.4 [ČŠ 2]: Let U be a region in $\mathbb{R}^k$, $<$ be a linear ordering of $\mathscr{D}(U)$ satisfying (4.3.6) and let the functions $f_1, \dots, f_n \in C^\infty(U, \mathbb{K})$ be locally linearly independent. We now describe how to determine a subregion $V \subseteq U$ and a complete n-tuple of derivatives $D_1, \dots, D_n \in \mathscr{D}(U)$ for the Wronski matrices (4.1.2) to satisfy the following two conditions:

(i) For each $p = 1, 2, \dots, n$,

$$\operatorname{rank} W_{p,n}[D_1, \dots, D_p; f_1, \dots, f_n](x) = p \quad (x \in V). \tag{4.3.7}$$

(ii) If $n > 1$ and $p \in \{2, 3, \dots, n\}$, then for each $D \in \mathscr{D}(U)$, $D < D_p$ implies that

$$\operatorname{rank} W_{p,n}[D_1, \dots, D_{p-1}, D; f_1, \dots, f_n](x) = p - 1 \quad (x \in V). \tag{4.3.8}$$

We start our recursion procedure by finding $x_0 \in U$ so that $f_i(x_0) \neq 0$ for some value of i. By continuity $\operatorname{rank} W_{1,n}[\mathrm{id}; f_1, \dots, f_n] = 1$ on some region V containing the point x_0. Assume now that for some $q \in \{1, 2, \dots, n-1\}$, a region $V \subseteq U$ and a complete q-tuple of derivatives $D_1, \dots, D_q \in \mathscr{D}(U)$ are chosen so that (4.3.7) and (4.3.8) are valid for each $p \in \{1, \dots, q\}$. There exists a derivative $D \in \mathscr{D}(U)$ satisfying

$$\operatorname{rank} W_{q+1,n}[D_1, \dots, D_q, D; f_1, \dots, f_n](x_0) = q+1 \quad \text{for some } x_0 \in V, \tag{4.3.9}$$

otherwise Theorem 4.1.1 would imply that each function f_i is a linear combination of a suitable q-tuple $f_{j_1}, \dots, f_{j_q}$ in the same subregion $W \subseteq V$, which would be a contradiction. (The region W and indices $j_1, \dots, j_q$ are selected so that

$$\det W_{q,q}[D_1, \dots, D_q; f_{j_1}, \dots, f_{j_q}](x) \neq 0 \quad \text{at each point } x \in W,$$

which is possible because of (4.3.7) with $p = q$.) Now we select D_{q+1} as (the unique) derivative satisfying (4.3.9) and being the smallest with respect to the ordering $<$. Then (4.3.7) and (4.3.8) hold automatically for $p = q+1$ (if necessary, the region V is restricted to a small neighbourhood of the point x_0 from (4.3.9) with $D = D_{q+1}$). Thus it remains to verify that the $(q+1)$-tuple $D_1, \dots, D_{q+1}$ is complete. Since $D_1, \dots, D_q$ is complete, we have only to show that $\hat{D} \in \{D_1, \dots, D_q\}$, for each derivative $\hat{D} \prec D_{q+1}$. Moreover, we can restrict our attention only to the derivatives $\hat{D}$ that precede D_{q+1} immediately, i.e. assume that $\hat{D}$ under consideration satisfies

$$D_{q+1} = \frac{\partial}{\partial x_s} \circ \hat{D} \quad \text{for some index } s \in \{1, \dots, k\}. \tag{4.3.10}$$

Suppose on the contrary, that such a $\hat{D}$ does not lie in $\{D_1, \dots, D_q\}$. Then the relation $\hat{D} < D_{q+1}$ implies that $D_{r-1} < \hat{D} < D_r$ for some $r \leq q+1$. Using now (4.3.7) with $p = r-1$ and (4.3.8) with $D = \hat{D}$ and $p = r$, we observe that the last row of the Wronski matrix $W_{r,n}[D_1, \dots, D_{r-1}, \hat{D}; f_1, \dots, f_n](x)$ is a linear combination of the previous ones, for each $x \in V$. Thus there exist some coefficient functions

$\gamma_i : V \to \mathbb{K}$ such that

$$\hat{D}\boldsymbol{f} = \sum_{i=1}^{r-1} \gamma_i \cdot D_i \boldsymbol{f} \qquad \text{on the region } V$$

where $\boldsymbol{f}$ denotes the row with entries f_1, ..., f_n. In view of (4.3.7) with $p = r-1$, the coefficients γ_i can be evaluated by means of Cramer's rule, which leads to the conclusion that $c_1, \ldots, c_{r-1} \in C^\infty(V, \mathbb{K})$. Consequently, we may differentiate the last row identity with respect to the variable x_s (recall that the index s has been defined in (4.3.10)). We obtain the identity

$$D_{q+1}\boldsymbol{f} = \sum_{i=1}^{r-1} \frac{\partial \gamma_i}{\partial x_s} \cdot D_i \boldsymbol{f} + \sum_{i=1}^{r-1} \gamma_i \cdot \left(\frac{\partial}{\partial x_s} \circ D_i \right) \boldsymbol{f} \quad \text{on the region } V. \tag{4.3.11}$$

Taking into account the second property in (4.3.6), we get

$$\frac{\partial}{\partial x_s} \circ D_i < \frac{\partial}{\partial x_s} \circ \hat{D} = D_{q+1} \quad (i = 1, \ldots, r-1)\,.$$

This along with (4.3.8) for $p = q+1$ implies that for each $x \in V$, any term in the second sum on the right-hand side of (4.3.11) is a linear combination of the q rows $D_1\boldsymbol{f}, \ldots, D_q\boldsymbol{f}$. Since in addition $r \leq q+1$, the whole right-hand side of (4.3.11) is a linear combination of the mentioned q-tuple, for each $x \in V$. This is in contradiction to (4.3.7) for $p = q+1$ and the proof of the completeness of $D_1, \ldots, D_{q+1}$ is finished.

Theorem 4.3.5 [ČŠ 2]: *Let $X \subseteq \mathbb{R}^k$ be a region and let the n-tuple of functions $f_1, \ldots, f_n \in C^\infty(X, \mathbb{K})$ be locally linearly independent. Then the set X_0 of all $x \in X$ satisfying the inequality* $\det W_n[D_i; f_j](x) \neq 0$, *for some complete n-tuple of derivatives $D_1, \ldots, D_n$ (which might differ in different subregions of X), is open and dense in X.*

Proof: Since X_0 is open (see the above proof of Theorem 4.3.1), it is sufficient to verify that X_0 is dense. Thus let $U \subseteq X$ be any subregion. Take a linear ordering $<$ of $\mathscr{D}(U)$ for (4.3.6) to be valid. Applying Algorithm 4.3.4, we find a complete n-tuple $D_1, \ldots, D_n \in \mathscr{D}(U)$ and a region $V \subseteq U$ such that, according to (4.3.7) with $p = n$, (4.3.3) holds for each $x \in V$. This means that $V \subseteq X_0$ which completes the proof of the density of X_0 in X. □

Remark 4.3.6: Note that the output of Algorithm 4.3.4, a complete n-tuple of derivatives $D_1, \ldots, D_n$, depends on the choice of the linear ordering $<$ with property (4.3.6). Considering the PDE system (4.2.2) that corresponds to this resulting n-tuple $D_1, \ldots, D_n$, we can easily see from formulas (4.2.3) and condition (4.3.8) that the coefficients a_{irs} in (4.2.2) possess the following natural property (related to the ordering $<$):

$$a_{irs}(x) = 0 \quad \text{for each } x \in V, \quad \text{whenever} \quad \frac{\partial}{\partial x_i} \circ D_r < D_s. \tag{4.3.12}$$

Our next result shows that for a given ordering $<$ of $\mathscr{D}(V)$, condition (4.3.12) is satisfied by at most one n-tuple of derivatives $D_1 = \mathrm{id} < \cdots < D_n$ (in the collection of all n-tuples of derivatives, not necessarily complete ones.) Consequently, under the restriction (4.3.12), there exists the unique PDE system (4.2.2) (in a sufficiently small subregion V) that (locally) describes a given n-dimensional space, possessing a locally linearly independent basis of smooth functions.

Theorem 4.3.7 [Ši 5]: *Let V be a region in $\mathbb{R}^k$ and let $<$ be a given linear ordering of $\mathscr{D}(V)$ that satisfies* (4.3.6). *Suppose that for a given n-tuple of functions $f_1, \ldots, f_n \in C^\infty(V, \mathbb{K})$, there exists an n-tuple of derivatives*

$$D_1 = \mathrm{id} < D_2 < \cdots < D_n \tag{4.3.13}$$

so that $\det W_n[D_i; f_j](x) \neq 0$ *for each $x \in V$ and that all the coefficients a_{irs} in* (4.2.3) *satisfy* (4.3.12). *Then such an n-tuple* (4.3.13) *is unique and complete.*

To make the proof of Theorem 4.3.7 more readable, we postpone it to Section 4.4, where some general properties of PDE systems (4.2.2) will be investigated. The significance of Theorem 4.3.7 is that it enables us to determine the unique "canonical" system in the class of all the PDE systems posessing the same n-dimensional solution space. As it will be shown in Theorem 4.4.2, such canonical systems can be reduced (i.e. substituted for by an equivalent subsystem of fewer equations), as demonstrated in part (iii) of Examples 4.2.1.

4.4. Generating and reducibility properties

To continue our discussion of PDE systems (4.2.2), let us retain all the assumptions and notations of Section 4.2.

A Generating Property: Note that besides (4.2.2), the functions f in the space S satisfy infinitely many other analogous equations. In fact, Theorem 4.1.1 implies that the identity $\det W_{n+1}[D_i, D; f_j, f] = 0$ holds on X, for *each* partial derivative $D \in \mathscr{D}(X)$. Expanding the above determinants as in Section 4.2, we obtain an infinite collection of equations on X:

$$Df = \sum_{s=1}^{n} a_{Ds}(x) \cdot D_s f \quad (f \in S,\ D \in \mathscr{D}(X)) \tag{4.4.1}$$

in which the coefficients $a_{Ds} \in C^\infty(X, \mathbb{K})$ are given by

$$a_{Ds} = \frac{\det W_n[D_1, \ldots, D_{s-1}, D, D_{s+1}, \ldots, D_n; f_j]}{\det W_n[D_i; f_j]} \tag{4.4.2}$$

and do not depend on the choice of the basis $f_1, \ldots, f_n$ of the space S. Let us emphasize that (4.2.2) is a finite subsystem of (4.4.1) selected according to Theorem 4.1.1 so that its solution space is identical with the original space S. Let us prove now that the subsystem (4.2.2) is a generator of the whole system (4.4.1) in the following sense.

Theorem 4.4.1 [Ši 5]: *Let X be a region in $\mathbb{R}^k$ and S be a linear space with a basis $f_1, \ldots, f_n \in C^\infty(X, \mathbb{K})$. Assume that the n-tuple $D_1 = \mathrm{id}, D_2, \ldots, D_n \in \mathscr{D}(X)$ is chosen so that* $\det W_n[D_i; f_j] \neq 0$ *on X. Then each equation in* (4.4.1) *can be obtained from the PDE system* (4.2.2) *by repeating the operations of differentiation and algebraic substitution. If in addition, a linear ordering $<$ of $\mathscr{D}(X)$ satisfies* (4.3.6) *and the coefficients a_{irs} of the generating system* (4.2.2) *possess the property* (4.3.12), *then also all the coefficients a_{Ds} in* (4.4.1) *do the same:*

$$a_{Ds}(x) = 0 \quad (x \in V), \quad \textit{whenever} \quad D < D_s\,. \tag{4.4.3}$$

Proof: We will proceed by induction with respect to the order of the derivative D from the left-hand side of (4.4.1). If $D = \mathrm{id}$, equation (4.4.1) is trivial. Assume now that for some D, equation (4.4.1) has been already obtained. Let us show how to derive equation (4.4.1) with D replaced by $\frac{\partial}{\partial x_i} \circ D$, for any $i = 1, \ldots, k$. Differentiating (4.4.1) with respect to x_i and substituting the derivatives $\frac{\partial}{\partial x_i} \circ D_r$ given by (4.2.2), we obtain

$$\begin{aligned}
\left(\frac{\partial}{\partial x_i} \circ D\right) f &= \frac{\partial}{\partial x_i}\left(\sum_{s=1}^{n} a_{Ds}(x) \cdot D_s f\right) \\
&= \sum_{s=1}^{n} \frac{\partial a_{Ds}}{\partial x_i}(x) \cdot D_s f + \sum_{r=1}^{n} a_{Dr}(x) \cdot \left(\frac{\partial}{\partial x_i} \circ D_r\right) f \\
&= \sum_{s=1}^{n} \frac{\partial a_{Ds}}{\partial x_i}(x) \cdot D_s f + \sum_{r=1}^{n} a_{Dr}(x) \cdot \left(\sum_{s=1}^{n} a_{irs}(x) \cdot D_s f\right) \\
&= \sum_{s=1}^{n} \left(\frac{\partial a_{Ds}}{\partial x_i}(x) + \sum_{r=1}^{n} a_{Dr}(x) \cdot a_{irs}(x)\right) \cdot D_s f\,.
\end{aligned} \tag{4.4.4}$$

Thus the generating property of (4.2.2) is established. To prove the second part of Theorem 4.4.1, assume now that the ordering $<$ is of type (4.3.6), that (4.3.12) holds and that for some given D, the coefficients a_{Ds} in (4.4.1) satisfy (4.4.3). In view of computations (4.4.4), it is sufficient only to verify that if some indices i and s satisfy

$$\frac{\partial}{\partial x_i} \circ D < D_s \tag{4.4.5}$$

then

$$\frac{\partial a_{Ds}}{\partial x_i}(x) + \sum_{r=1}^{n} a_{Dr}(x) \cdot a_{irs}(x) = 0 \quad (x \in V). \tag{4.4.6}$$

The first term in the left-hand side vanishes on V, because (4.4.5) implies that $D < D_s$ and hence $a_{Ds} = 0$ on V by (4.4.3). For the same reason, the r-th term of the sum in (4.4.6) vanishes on V, whenever r satisfies $D < D_r$. On the other hand, if $D \geq D_r$, then

$$\frac{\partial}{\partial x_i} \circ D_r \leq \frac{\partial}{\partial x_i} \circ D < D_s$$

and thus $a_{irs} = 0$ on V by (4.3.12). This completes the proof of (4.4.6). □

Proof of Theorem 4.3.7: Suppose, on the contrary, that besides (4.3.13), there exists another n-tuple of derivatives $\hat{D}_1 = \mathrm{id} < \cdots < \hat{D}_n$ with the same properties as in the statement of Theorem 4.3.7. Take the index $p > 1$ so that $D_p \neq \hat{D}_p$, while $D_i = \hat{D}_i$ for each $i < p$. It is no loss of generality to assume $\hat{D}_p < D_p$ (interchange both the n-tuple if not). Let S denotes the linear space spanned by the n-tuple of functions $f_1, \ldots, f_n$. For this space S, we can consider equation (4.4.1) with $D = \hat{D}_p$:

$$\hat{D}_p f = \sum_{s=1}^{n} a_{*s}(x) \cdot D_s f \quad (x \in V,\, f \in S)$$

where $* = \hat{D}_p$. Since $\hat{D}_p < D_s$ for each $s \geq p$, Theorem 4.4.1 implies that all a_{*s} vanish on V, for such indices s. Hence the last equation can be simplified to

$$\hat{D}_p f = \sum_{s=1}^{p-1} a_{*s}(x) \cdot D_s f = \sum_{s=1}^{p-1} a_{*s}(x) \cdot \hat{D}_s f$$

which means that the first p rows of the nonsingular matrix $W_n[\hat{D}_i; f_j](x)$ are linearly dependent for each $x \in V$, which is a contradiction. The uniqueness of the n-tuple (4.3.13) is proved. Now we verify that the n-tuple (4.3.13) satisfying (4.3.12) is a complete one. Suppose on the contrary the existence of indices i and p such that

$$D_p = \frac{\partial}{\partial x_i} \circ D \quad \text{and} \quad D \notin \{D_1, \ldots, D_{p-1}\}.$$

Since $D \prec D_p$ (and thus $D < D_p$), there exists an index $q<p$ such that $D_q < D < D_{q+1}$. To obtain (4.4.1) for $D=D_p$, we utilize (4.4.4). By virtue of Theorem 4.4.1, the coefficient a_{Ds} vanishes on V, for each $s > q$. Moreover, if $r \le q$, then

$$\frac{\partial}{\partial x_i} \circ D_r \le \frac{\partial}{\partial x_i} \circ D_q < \frac{\partial}{\partial x_i} \circ D = D_p,$$

and hence (4.3.12) ensures $a_{irs} = 0$ on V $(1 \le r \le q,\, p \le s \le n)$. Consequently, equation (4.4.4) with $D=D_p$ can be written on V as follows

$$D_p f = \left(\frac{\partial}{\partial x_i} \circ D\right) f = \sum_{s=1}^{q} \frac{\partial a_{Ds}}{\partial x_i}(x) D_s f + \sum_{s=1}^{p-1} \left(\sum_{r=1}^{q} a_{Dr}(x) \cdot a_{irs}(x)\right) \cdot D_s f\,.$$

This along with $q < p$ leads to the conclusion that the p-th row of the nonsingular Wronski matrix $W_n[D_i; f_j](x)$ is a linear combination of the previous ones, for each $x \in V$. This contradiction proves the completeness of (4.3.13). □

A Reducibility Property: Let us return to a PDE system (4.2.2) with a given n-dimensional solution space S. In Theorem 4.4.1, we have shown that such a system enables us to compute all the differential equations (4.4.1). Thus an interesting problem immediately arises: when does there exist a subsystem of (4.2.2) possessing the same generating property (as well as in part (iii) of Examples 4.2.1)? First of all, as explained in the introductory part of Section 4.2, we may exclude those equations in (4.2.2) that are trivial. (Recall that the trivial equations in (4.2.2) correspond exactly to those pairs of indices for which the operator $\frac{\partial}{\partial x_i} \circ D_r$ belongs to the n-tuple $D_1, D_2, \ldots, D_n$. More generally, an equation in (4.4.1) is trivial if and only if D lies in $D_1, \ldots, D_n$.) On the other hand, it is easily understood that each generating subsystem of (4.2.2)

$$Df = \sum_{s=1}^{n} a_{Ds}(x) \cdot D_s f \quad (f \in S,\ D \in \mathscr{G}_0) \tag{4.4.7}$$

has to include equations for all the derivatives D that are *minimal* elements of the (nonempty) collection

$$\mathscr{G} = \left\{\frac{\partial}{\partial x_i} \circ D_r \;\middle|\; i=1,\ldots,k,\ r=1,\ldots,n\right\} \setminus \{D_1, \ldots, D_n\} \tag{4.4.8}$$

in the sense of partial ordering $\prec$. (The collection $\mathscr{G}$ corresponds to all the nontrivial equations in (4.2.2).) The following theorem shows that such an *a priori* minimal subsystem (4.4.7), when $\mathscr{G}_0$ is the set of the minimal elements of $(\mathscr{G}, \prec)$, possesses the desired generating property, provided that the derivatives $D_1, \ldots, D_n$ are selected in the "canonical" way as desribed in Theorem 4.3.7.

Theorem 4.4.2 [Ši 5]: *Let V be a region in $\mathbb{R}^k$ and let S be a linear functional space with a basis $f_1, \ldots, f_n \in C^\infty(V, \mathbb{K})$. Assume also that an n-tuple of derivations $D_1, \ldots, D_n \in \mathscr{D}(V)$ satisfies* $\det W_n[D_i; f_j](x) \neq 0$ *at each point $x \in V$ and that the set $\mathscr{G}_0$ of all the minimal elements of the set $(\mathscr{G}, \prec)$ defined in* (4.4.8) *possesses the following property: there exists a linear ordering $(\mathscr{D}(V), <)$ of type* (4.3.6) *such that for each $D \in \mathscr{G}_0$, the coefficients a_{Ds} defined by* (4.4.2) *satisfy* (4.4.3). *Then each equation*

$$Df = \sum_{s=1}^{n} a_{Ds}(x) \cdot D_s f \quad (f \in S), \tag{4.4.9}$$

for any derivative $D \in \mathscr{D}(V)$, can be obtained from the system (4.4.7) *by repeating the operations of differentiation and algebraic substitution. Moreover, all the coefficients a_{Ds} in* (4.4.9) *possess the property* (4.4.3).

Proof: In view of the generating property stated in Theorem 4.4.1 we need to prove only the conclusions concerning equations (4.4.9) with $D \in \mathscr{G}$. To this purpose, we will proceed by induction on the linearly ordered and countable set $(\mathscr{D}(V), <)$ (see a remark in the last paragraph of this proof).

Since the smallest element D_* in $(\mathscr{G}, <)$ must be minimal in $(\mathscr{G}, \prec)$, equation (4.4.9) with $D = D_*$ is involved in (4.4.7). Assume now that $\hat{D} \in \mathscr{G}$ is a (fixed) derivative such that equation (4.4.9) has already obtained and that its coefficients a_{Ds} satisfy (4.4.3), for each derivative $D < \hat{D}$. We will show how to obtain equation (4.4.9) with $D = \hat{D}$.

We can restrict our attention only to the case, when $\hat{D} \notin \mathscr{G}_0$, that is $\tilde{D} \prec \hat{D}$, for some $\tilde{D} \in \mathscr{G}$. Then $\hat{D} = \tilde{D} \circ D'$, where $D' \neq \mathrm{id}$ and hence $\frac{\partial}{\partial x_i} \preceq D'$, for some index $i \in \{1, 2, \ldots, k\}$. (As usual, $D'' \preceq D'$ means that either $D'' = D'$ or $D'' \prec D'$.) Let us define the derivative D'' by $D' = \frac{\partial}{\partial x_i} \circ D''$ and verify that equation (4.4.9) with $D = \hat{D}$ can be obtained in the case when

$$\tilde{D} \circ D'' \in \mathscr{G}. \tag{4.4.10}$$

In fact, under the condition (4.4.10), equation (4.4.9) with $D = \tilde{D} \circ D''$ is assumed to be already derived by induction (because of $\tilde{D} \circ D'' < \hat{D}$). Differentiating this equation with respect to x_i, we obtain

$$\begin{aligned} \hat{D}f &= \frac{\partial}{\partial x_i} \circ \left(\tilde{D} \circ D'' f\right) = \frac{\partial}{\partial x_i}\left(\sum_{s \in I} a_{*s}(x) \cdot D_s f\right) \\ &= \sum_{s \in I} \frac{\partial a_{*s}}{\partial x_i}(x) \cdot D_s f + \sum_{s \in I} a_{*s}(x) \cdot \left(\frac{\partial}{\partial x_i} \circ D_s\right) f \end{aligned} \tag{4.4.11}$$

where $I = \{s \mid D_s < \tilde{D} \circ D''\}$, because of property (4.4.3). To obtain the desired equation (4.4.9) for $D = \hat{D}$, it is sufficient to substitute each operator in the last sum of (4.4.11) by the sum from the corresponding equation (4.4.9). This is possible, because each of the derivatives $\frac{\partial}{\partial x_i} \circ D_s$ $(s \in I)$ either lies in $\{D_j \mid D_j < \hat{D}\}$, or it is an element of $\mathscr{G}$ smaller than $\hat{D}$; in the last case equation (4.4.9) with $D = \frac{\partial}{\partial x_i} \circ D_s$ is at our disposal by the induction assumption. Since the coefficients of all the equations substituted into (4.4.11) satisfy (4.4.3), the coefficients of the resulting equation for $D = \hat{D}$ do so as well.

It remains to prove that (4.4.10) really holds. By definition of $\mathscr{G}$, the relation $\hat{D} \in \mathscr{G}$ ensures the existence of indices j and p satisfying

$$\hat{D} = \frac{\partial}{\partial x_j} \circ D_p \,. \tag{4.4.12}$$

Since $\tilde{D} \in \mathscr{G}$, $\tilde{D}$ is not present in the complete n-tuple $D_1, \ldots, D_n$. Hence $\tilde{D} \prec D_p$ is impossible, even though $\tilde{D} \prec \hat{D}$. Comparing this with (4.4.12), we conclude that $\frac{\partial}{\partial x_j} \prec \tilde{D}$ must be true. Then (4.4.12), along with $\hat{D} = \tilde{D} \circ D'$, implies that $D' \prec D_p$. Due to the completeness of the p-tuple $D_1, \ldots, D_p$, there exists an index $q < p$ such that $D_p = D_q \circ D'$. From the following series of equalities

$$\begin{aligned} \tilde{D} \circ D'' \circ \frac{\partial}{\partial x_i} &= \tilde{D} \circ D' = \hat{D} = \frac{\partial}{\partial x_j} \circ D_p \\ &= \frac{\partial}{\partial x_j} \circ D_q \circ D' = \frac{\partial}{\partial x_j} \circ D_q \circ D'' \circ \frac{\partial}{\partial x_i} \end{aligned}$$

we can see that $\tilde{D} \circ D'' = \frac{\partial}{\partial x_j} \circ (D_q \circ D'')$. Moreover, $D_q \circ D'' \prec D_q \circ D' = D_p$ and thus the derivative $D_q \circ D''$ lies in the complete p-tuple $D_1, \ldots, D_p$. This means that $\tilde{D} \circ D''$ lies in the first of the sets in the right-hand side of (4.4.8). On the other hand, it is clear that the derivative $\tilde{D} \circ D''$ cannot lie in the complete n-tuple $D_1, \ldots, D_n$, otherwise the derivative $\tilde{D} \preceq \tilde{D} \circ D''$ should do so as well, which would be in contradiction to $\tilde{D} \in \mathscr{G}$. This proves (4.4.10).

To complete the proof by induction on $(\mathscr{D}(V), <)$, let us note that this ordered set is not (in general) isomorphic with the ordered set of all positive integers $(\mathbb{N}, <)$ (see the example of the lexicographical ordering at the end of Definitions 4.3.3). For the above proof by induction to be correct, we therefore have to show that under the condition (4.3.6), the ordered set $(\mathscr{G}, <)$ does not include any infinite decreasing sequence $D_1 > D_2 > D_3 > \cdots$. However, this property easily follows from the fact that any infinite sequence of derivatives $D_1, D_2, \cdots \in \mathscr{D}(V)$ contains an infinite subsequence $D_{j_1} \preceq D_{j_2} \preceq \cdots$ (with $j_1 < j_2 < \cdots$). (Consider the representations

$$D_j = \frac{\partial^{\alpha_{1j} + \alpha_{2j} + \cdots + \alpha_{kj}}}{\partial x_1^{\alpha_{1j}} \partial x_2^{\alpha_{2j}} \ldots \partial x_k^{\alpha_{kj}}} \tag{4.4.13}$$

and utilize repeatedly the fact that each infinite sequence of nonnegative integers includes an infinite nondecreasing subsequence.) □

Remark 4.4.3: Although no equation in the reduced system (4.4.7) can be obtained from the other ones (by repeating the operations of differentiation and algebraic substitution), we may not assert that the coefficients a_{Ds} are mutually *independent*. This can be easily seen even in the case $n=1$, when a PDE system of k equations

$$\frac{\partial f}{\partial x_i} = a_i(x_1, \dots, x_k) f \quad (i=1,2,\dots,k)$$

has a solution space of dimension 1. In fact, the coefficients a_i have to satisfy the $\binom{k}{2}$ conditions

$$\frac{\partial a_i}{\partial x_j} = \frac{\partial a_j}{\partial x_i} \quad (1 \le i < j \le k)$$

following from the representation of the mixed derivatives

$$\frac{\partial^2 f}{\partial x_i \partial x_j} = \frac{\partial}{\partial x_j}(a_i f) = \frac{\partial a_i}{\partial x_j} f + a_i \frac{\partial f}{\partial x_j} = \left(\frac{\partial a_i}{\partial x_j} + a_i a_j \right) f \,.$$

5 DECOMPOSITIONS OF SMOOTH FUNCTIONS ON MANIFOLDS

In this chapter, we present necessary and sufficient conditions for a smooth function h in two *vector* variables x, y to be represented in the form

$$h(x,y) = \sum_{i=1}^{n} f_i(x) g_i(y) \,. \tag{5a}$$

Although the question of the decomposition of a function of $p + q$ real variables

$$h(x_1, \ldots, x_{p+q}) = \sum_{i=1}^{n} f_i(x_1, \ldots, x_p) g_i(x_{p+1}, \ldots, x_{p+q})$$

may be of some interest in itself, we recall the crucial role it plays for the solution of the fundamental problem for functions in several variables (see the reduction method in Chapter 3, Section 3.1).

Necessary and sufficient conditions for the decomposition (5a) will be given in Section 5.1 by using a new notion of the general Wronski matrix of the function h

$$W_n[D_i; d_j]h := \begin{pmatrix} D_1 \circ d_1 h & D_1 \circ d_2 h & \ldots & D_1 \circ d_n h \\ D_2 \circ d_1 h & D_2 \circ d_2 h & \ldots & D_2 \circ d_n h \\ \vdots & \vdots & \ddots & \vdots \\ D_n \circ d_1 h & D_n \circ d_2 h & \ldots & D_n \circ d_n h \end{pmatrix} \qquad (n=1,2,\ldots). \tag{5b}$$

We will assume that $h \in C^\infty(X \times Y, \mathbb{K})$, where X and Y are regions in $\mathbb{R}^p$ and $\mathbb{R}^q$ respectively, and that $D_1, D_2, \ldots$ and $d_1, d_2, \ldots$ are arbitrary partial derivatives from the sets $\mathscr{D}(X)$ and $\mathscr{D}(Y)$, respectively (for this notation, see Section 4.1).

We will also deal with the global decomposition problem (5a) for functions h defined on the Cartesian product of two topological manifolds. The approach we are following is that of gluing together the local decompositions (5a) of the function h under consideration.

5.1. Differential conditions

The basic result Theorem 2.1.1 deals with the decomposition problem (5a) in the simplest case, when the independent variables x and y are scalar ones. The goal of this section is to generalize Theorem 2.1.1 to the case of vector variables in two steps. First we prove that the vanishing of each $(n+1)$-th order general Wronski matrix of h is a *necessary* condition for h to be of the form (5a), with a given number n of terms $f_i g_i$.

Theorem 5.1.1 [ČŠ 2]: *Let $X \subseteq \mathbb{R}^p$ and $Y \subseteq \mathbb{R}^q$ be two regions and let a function $h \in C^\infty(X \times Y, \mathbb{K})$ be of the form* (5a) *on $X \times Y$, where $f_1, \ldots, f_n \in C^\infty(X, \mathbb{K})$ and $g_1, \ldots, g_n \in C^\infty(Y, \mathbb{K})$. Then for two arbitrary $(n+1)$-tuples of derivatives*

$$D_1, \ldots, D_{n+1} \in \mathscr{D}(X) \quad \text{and} \quad d_1, \ldots, d_{n+1} \in \mathscr{D}(Y),$$

the following identity holds

$$\det W_{n+1}[D_i; d_j]h(x, y) = 0 \quad (x \in X,\ y \in Y)\,. \tag{5.1.1}$$

If in addition each of the n-tuples $f_1, \ldots, f_n$ and $g_1, \ldots, g_n$ is locally linearly independent (see Section 1.1)*, then the set of all pairs $(x, y) \in X \times Y$, for which there exist two complete n-tuples of derivatives $D_1, \ldots, D_n \in \mathscr{D}(X)$ and $d_1, \ldots, d_n \in \mathscr{D}(Y)$ with the property*

$$\det W_n[D_i; d_j]h(x, y) \neq 0\,, \tag{5.1.2}$$

is open and dense in $X \times Y$.

Proof: For every given $y \in Y$, the matrix $W_{n+1}[D_i; d_j]h(-, y)$ is the Wronski matrix (4.1.2) of the system of $n+1$ functions

$$d_1 h(-, y),\ d_2 h(-, y), \ldots, d_{n+1} h(-, y)$$

in the variable x. Since the function h is of the form (5a), the above system is linearly dependent in the set X. Consequently, the identity (5.1.1) follows from part (i) of Theorem 4.1.1. The representation (5a) also implies that the Wronski matrix $W_n[D_i; d_j]h$ is a matrix product of the Wronski matrix $W_n[D_i; f_j]$ and the transpose of the Wronski matrix $W_n[d_i; g_j]$ (see Section 4.1, where this type of matrix was defined). Thus in the case of the locally linearly independent n-tuples $f_1, \ldots, f_n$ and $g_1, \ldots, g_n$, the conclusion for the set of all pairs (x, y) with the property (5.1.2) follows from Theorem 4.3.5. □

We now state some *sufficient* conditions for h to be of the form (5a), which are nontrivial extensions of that of the second part of Theorem 2.1.1.

Theorem 5.1.2 [ČŠ 2]: *Let $X \subseteq \mathbb{R}^p$ and $Y \subseteq \mathbb{R}^q$ be two regions and let h be a function in $C^\infty(X \times Y, \mathbb{K})$, which possesses the following property: There exist two n-tuples of derivatives*

$$D_1 = \mathrm{id}, D_2, \ldots, D_n \in \mathscr{D}(X) \quad \textit{and} \quad d_1 = \mathrm{id}, d_2, \ldots, d_n \in \mathscr{D}(Y)$$

such that

$$\det W_n[D_i; d_j]h(x,y) \neq 0 \quad (x \in X,\ y \in Y) \tag{5.1.3}$$

and that for each first-order derivatives $D \in \mathscr{D}^{(1)}(X)$ and $d \in \mathscr{D}^{(1)}(Y)$ the following identities

$$\det W_{n+1}[D_1, \ldots, D_n, D \circ D_r; d_1, \ldots, d_n, d \circ d_s]h(x,y) = 0 \quad (x \in X,\ y \in Y) \tag{5.1.4}$$

hold for all $r, s \in \{1, 2, \ldots, n\}$. Then the function h has a decomposition (5a) *on the set $X \times Y$, with locally linearly independent components $f_i \in C^\infty(X, \mathbb{K})$ and $g_i \in C^\infty(Y, \mathbb{K})$. Moreover, the function h can be expressed as a matrix product of three matrices of sizes $1 \times n$, $n \times n$ and $n \times 1$:*

$$\begin{aligned} h(x,y) = \big(h(x,y_0), d_2h(x,y_0), \ldots, d_nh(x,y_0)\big) & \\ \times W_n^{-1}[D_i; d_j]h(x_0,y_0) \cdot \begin{pmatrix} h(x_0,y) \\ D_2h(x_0,y) \\ \vdots \\ D_nh(x_0,y) \end{pmatrix} & \end{aligned} \tag{5.1.5}$$

where (x_0, y_0) is an arbitrarily chosen point of the product $X \times Y$.

Proof: Let $D \in \mathscr{D}^{(1)}(X)$, $d \in \mathscr{D}^{(1)}(Y)$ and $r, s \in \{1, 2, \ldots, n\}$. According to (5.1.3) and (5.1.4) for all $x \in X$ and $y \in Y$, the $(n+1)$-th row of the matrix (5.1.4) is a linear combination of the first n rows. Thus there exist functions $a_{Ddrsi} = a_{Ddrsi}(x,y)$ such that the following $n+1$ identities

$$D \circ D_r \circ d_j h = \sum_{i=1}^{n} a_{Ddrsi} \cdot D_i \circ d_j h \quad (j = 1, 2, \ldots, n) \tag{5.1.6}$$

and

$$D \circ D_r d \circ d_s h = \sum_{i=1}^{n} a_{Ddrsi} \cdot D_i \circ d \circ d_s h \tag{5.1.7}$$

are satisfied on $X \times Y$. The coefficients a_{Ddrsi} can be evaluated from the n equations (5.1.6) using Cramer's rule. In this way we verify that each function a_{Ddrsi} is of class C^∞ on $X \times Y$ and it does not depend upon the choice of $d \in \mathscr{D}^{(1)}(Y)$ as well as upon the choice of the index $s \in \{1, 2, \dots, n\}$. Thus we can write $a_{Ddrsi} = a_{Dri}$. Applying the operator $d \in \mathscr{D}^{(1)}(Y)$ to the identity (5.1.6) for a given j and then subtracting the identity (5.1.7) with $s = j$, we obtain

$$0 = \sum_{i=1}^{n} da_{Dri} \cdot D_i \circ d_j h \quad (j = 1, 2, \dots, n).$$

Taking in account (5.1.3), the last identity means that $da_{Dri}(x, y) = 0$ for each pair $(x, y) \in X \times Y$ and for each $d \in \mathscr{D}^{(1)}(Y)$. Thus the function a_{Dri} does not depend upon the variable $y \in Y$. This is why we can write $a_{Dri} = a_{Dri}(x)$. Since $d_1 = \mathrm{id}$, it follows from (5.1.6) with $j = 1$ that for each $y \in Y$, the function $h(-, y)$ is a solution of a system of partial differential equations

$$D \circ D_r f = \sum_{i=1}^{n} a_{Dri}(x) \cdot D_i f \tag{5.1.8}$$

where $D \in \mathscr{D}^{(1)}(X)$ and $r \in \{1, 2, \dots, n\}$. However, the last system is a PDE system (4.2.2) and according to (5.1.6), it has an n-tuple of solutions

$$d_1 h(-, y_0), d_2 h(-, y_0), \dots, d_n h(-, y_0) \tag{5.1.9}$$

where y_0 is an arbitrarily chosen point in Y. Using (5.1.3) with $y = y_0$ and applying part (ii) of Theorem 4.1.1, we may conclude that the function $h(-, y)$ is a linear combination of the n-tuple (5.1.9) in the set X. This means that there exist functions $g_j : Y \to \mathbb{K}$ for which the equality

$$h(x, y) = \sum_{j=1}^{n} g_j(y) \cdot d_j h(x, y_0) \tag{5.1.10}$$

holds at each point $(x, y) \in X \times Y$. Consequently, the function h is of the form (5a), where $f_j = d_j h(-, y_0)$. To compute the components g_j, one can apply the operator D_i to the identity (5.1.10) and then make the substitution $x = x_0$ to get

$$D_i h(x_0, y) = \sum_{j=1}^{n} g_j(y) \cdot D_i \circ d_j h(x_0, y_0) \quad (i = 1, 2, \dots, n).$$

This can also be written in the matrix product form

$$\begin{pmatrix} D_1h(x_0,y) \\ D_2h(x_0,y) \\ \vdots \\ D_nh_(x_0,y) \end{pmatrix} = W_n[D_i;d_j]h(x_0,y_0) \cdot \begin{pmatrix} g_1(y) \\ g_2(y) \\ \vdots \\ g_n(y) \end{pmatrix}.$$

Multiplying this by $W_n^{-1}[D_i;d_j]h(x_0,y_0)$ from the left-hand side, we obtain the values of $g_1(y),\dots,g_n(y)$. Substituting these values into (5.1.10), we derive the representation formula (5.1.5). □

5.2. Global decompositions

In both theorems of Section 5.1, it is assumed that the independent variables x and y have coordinate representations $x_1,x_2,\dots,x_n$ and $y_1,y_2,\dots,y_n$, respectively. Thus in the case when a given function $h = h(x,y)$ is defined on the Cartesian product of two topological manifolds $X \times Y$, one can apply these results only *locally* in the following sense. Given a pair $(x,y) \in X \times Y$, there exists a neighbourhood $U \subseteq X$ of the point x with coordinates $x_1,x_2,\dots,x_n$ and a neighbourhood $V \subseteq Y$ of the point y with coordinates $y_1,y_2,\dots,y_n$. Then $U \times V$ is such a neigbourhood of the given pair (x,y) in $X \times Y$ in which

$$h(x,y) = h(x_1,\dots,x_p,y_1,\dots,y_q)\,.$$

Hence we can apply Theorems 5.1.1 and 5.1.2 to the restriction $h|(U \times V)$. Now, the following important question arises: Can such local decompositions (5a) *be glued* together into a *global* decomposition on the whole product $X \times Y$? The next result gives a positive answer to this question for the case when each local decomposition (5a) that is minimal (see Section 2.3) has the same number n of terms f_ig_i. This result can be stated in a more general situation when the given function h (not necessarily continuous) is defined on the Cartesian product of two arcwise connected topological spaces.

Theorem 5.2.1 [ČŠ 2]: *Let X and Y be two arcwise connected topological spaces. Suppose that a function $h : X \times Y \to \mathbb{K}$ has the following property: Each point $(x_0,y_0) \in X \times Y$ has such a neighbourhood $U \times V \subseteq X \times Y$ on which*

$$h(x,y) = \sum_{i=1}^{n} f_i(x)g_i(y) \tag{5.2.1}$$

where $n \geq 1$ is a given integer (independent of the choice of the point (x_0,y_0)) and $f_i : U \to \mathbb{K}$ and $g_i : V \to \mathbb{K}$ are locally linearly independent functions. Then the function h has a decomposition (5.2.1) *on the whole product $X \times Y$.*

For the proof of the theorem we need the following:

Lemma 5.2.2: *Let X and Y be two arcwise connected topological spaces and let $\gamma_1 : [0,1] \to X$ and $\gamma_2 : [0,1] \to Y$ be two continuous mappings, i.e. curves. Denote by*

$$\Gamma_i = \{\gamma_i(t) \mid t \in [0,1]\} \quad (i=1,2) \tag{5.2.2}$$

and suppose that a given function $h : X \times Y \to \mathbb{K}$ has the following property: there exists an integer $n \geq 1$ such that each point $(x,y) \in X \times Y$ has a neighbourhood $\tilde{U} \times \tilde{V} \subseteq X \times Y$ on which h permits a representation (5.2.1), *with some n-tuples of locally linearly independent functions $f_i : \tilde{U} \to \mathbb{K}$ and $g_i : \tilde{V} \to \mathbb{K}$. Then there exist open sets $\Gamma_1 \subseteq U \subseteq X$ and $\Gamma_2 \subseteq V \subseteq Y$ such that the function h is of the form* (5.2.1) *on $U \times V$, with some n-tuples of functions $f_i : U \to \mathbb{K}$ and $g_i : V \to \mathbb{K}$.*

Proof of Lemma 5.2.2: First we outline the main steps of the proof. To avoid unnecessary repetition, we will assume that the letters U and V will always denote nonempty open sets in topological spaces X and Y respectively, while the bold letters $\boldsymbol{f}$ and $\boldsymbol{g}$ will stand for n-tuples of locally independent functions $f_1, \dots, f_n$ and $g_1, \dots, g_n$, respectively. Since these n-tuples may be assumed to be columns, i.e. matrices of size $n \times 1$, the right-hand side of (5.2.1) can be written as a matrix product $\boldsymbol{f}^T(x)$ *times* $\boldsymbol{g}(y)$, where ^{T} denotes transposition. The main idea of the proof consists in the *extension* of decompositions (5.2.1) along the curves. Thus besides the "whole" curves Γ_i in (5.2.2), we will also consider the "shorted" curves

$$\Gamma_i(t) = \{\gamma_i(s) \mid s \in [0,t]\} \quad \text{and} \quad \Gamma_i(s,t) = \{\gamma_i(\tau) \mid \tau \in [s,t]\} \quad (i=1,2).$$

Throughout the proof, assume that the conditions of Lemma 5.2.2 are satisfied. The first step is to prove the following proposition:

(i) *There exist sets U_0, V_0 and mappings $\boldsymbol{f}$, $\boldsymbol{g}_0$ such that $\Gamma_1 \subseteq U_0$, $\gamma_2(0) \in V_0$ and the equality $h = \boldsymbol{f}^T \boldsymbol{g}_0$ holds on the set $U_0 \times V_0$.*

Let us fix the set U_0 and the mapping $\boldsymbol{f}$ from (i) and consider now the set P of all reals $t \in [0,1]$ possessing the following property: there are sets U_t, V_t and a mapping $\boldsymbol{g}_t$ such that $\Gamma_1 \subseteq U_t \subseteq U_0$, $\Gamma_2(t) \subseteq V_t$ and $h = \boldsymbol{f}^T \boldsymbol{g}_t$ on the set $U_t \times V_t$. Then (i) means that $0 \in P$. Moreover, P is evidently open in $[0,1]$ and it holds that $[0,t] \subseteq P$ whenever $t \in P$. Since the conclusion of Lemma 5.2.2 is equivalent to $1 \in P$, it is sufficient to verify only the implication $[0,t) \subseteq P \Rightarrow t \in P$, for each $t \leq 1$. So asssume that $[0,t) \subseteq P$ for some given t and consider the set R of all reals $\tau \in [0,1]$ with the property: there are sets $\hat{U}_\tau$, $\hat{V}_\tau$ and a mapping $\hat{\boldsymbol{g}}_\tau$ such that $\Gamma_1(\tau) \subseteq \hat{U}_\tau \subseteq U_0$, $\Gamma_2(t) \subseteq \hat{V}_\tau$ and $h = \boldsymbol{f}^T \hat{\boldsymbol{g}}_\tau$ on the set $\hat{U}_\tau \times \hat{V}_\tau$. We need only to show that $1 \in R$, because then we can set $U_t = \hat{U}_1$, $V_t = \hat{V}_1$, $\boldsymbol{g}_t = \hat{\boldsymbol{g}}_1$ and thus conclude that $t \in P$. Since the set R is evidently open in $[0,1]$ and since $[0,\tau] \subseteq R$ whenever $\tau \in R$, it only suffices yet to prove the two properties:

(ii) $0 \in R$,

(iii) *if* $[0,\tau) \subseteq R$, *then* $\tau \in R$, *for each* $\tau \le 1$.

In the rest of the proof, we verify our claims (i), (ii) and (iii).

Verification of (i). In view of the conditions of Lemma 5.2.2, there exists a neighbourhood $U'_0 \times V'_0$ of the point $(\gamma_1(0), \gamma_2(0))$ on which $h = \boldsymbol{f}^T \boldsymbol{g}$ holds, with suitable $\boldsymbol{f}$ and $\boldsymbol{g}$. Let us fix such a $\boldsymbol{g}$ (defined on a given V'_0) and define the set Q of all reals $\tau \in [0,1]$ with the following property: there are sets U'_τ, V'_τ and a mapping $\boldsymbol{f}_\tau$ such that $h = \boldsymbol{f}_\tau^T \boldsymbol{g}$ on the set $U'_\tau \times V'_\tau$, $\Gamma_1(\tau) \subseteq U'_\tau$ and $\gamma_2(0) \in V'_\tau \subseteq V'_0$. It is clear that Q is open in $[0,1]$, $0 \in Q$ and that $[0,\tau] \subseteq Q$ whenever $\tau \in Q$. Since our goal in (i) is to show that $1 \in Q$, it suffices only to verify that $\tau \in Q$ whenever τ satisfies $[0,\tau) \subseteq Q$. Having chosen such a τ, we first find a neighbourhood $\tilde{U} \times \tilde{V}$ of the point $(\gamma_1(\tau), \gamma_2(0))$ on which we have $h = \tilde{\boldsymbol{f}}^T \tilde{\boldsymbol{g}}$, with suitable $\tilde{\boldsymbol{f}}$ and $\tilde{\boldsymbol{g}}$. Since $\gamma_1(\sigma) \to \gamma_1(\tau)$ as $\sigma \to \tau$, it holds that $\Gamma_1(\sigma,\tau) \subseteq \tilde{U}$ for some $\sigma < \tau$. It follows from $\sigma \in Q$ that $h = \boldsymbol{f}_\sigma^T \boldsymbol{g}$ on the set $U'_\sigma \times V'_\sigma$, described in the definition of Q above. In this way, we get a "double" representation $h = \tilde{\boldsymbol{f}}^T \tilde{\boldsymbol{g}} = \boldsymbol{f}_\sigma^T \boldsymbol{g}$ on the set $(\tilde{U} \cap U'_\sigma) \times (\tilde{V} \cap V'_\sigma)$, which is an (open) neighbourhood of the point $(\gamma_1(\sigma), \gamma_2(0))$. By Theorem 2.3.1, the last implication is possible only if $\boldsymbol{f}_\sigma = C^T \tilde{\boldsymbol{f}}$ on $\tilde{U} \cap U'_\sigma$ and $\boldsymbol{g} = C^{-1} \tilde{\boldsymbol{g}}$ on $\tilde{V} \cap V'_\sigma$, for some constant nonsingular matrix C of size $n \times n$. Consequently, we may put $U'_\tau = \tilde{U} \cup U'_\sigma$, $V'_\tau = \tilde{V} \cap V'_\sigma$ and define (in a correct way) the mapping

$$\boldsymbol{f}_\tau(x) := \begin{cases} \boldsymbol{f}_\sigma(x) & \text{for } x \in U'_\sigma \\ C^T \tilde{\boldsymbol{f}}(x) & \text{for } x \in \tilde{U} \end{cases}$$

on the set U'_τ. Then $\Gamma_1(\tau) = \Gamma_1(\sigma) \cup \Gamma_1(\sigma,\tau) \subseteq U'_\tau$, $\gamma_2(0) \in V'_\tau \subseteq V_0$ and the set $U'_\tau \times V'_\tau$ is evidently a subset of $U'_\sigma \times V'_\sigma \cup [\tilde{U} \times (\tilde{V} \cap V'_\sigma)]$. Moreover, we have $h = \boldsymbol{f}_\sigma^T \boldsymbol{g} = \boldsymbol{f}_\tau^T \boldsymbol{g}$ on the set $U'_\sigma \times V'_\sigma$ and $\boldsymbol{f}_\tau^T \boldsymbol{g} = (C^T \tilde{\boldsymbol{f}})^T (C^{-1} \tilde{\boldsymbol{g}}) = \tilde{\boldsymbol{f}}^T \tilde{\boldsymbol{g}} = h$ on the set $\tilde{U} \times (\tilde{V} \cap V'_\sigma)$. Thus $h = \boldsymbol{f}_\tau^T \boldsymbol{g}$ on the set $U'_\tau \times V'_\tau$, which proves $\tau \in Q$.

Verification of (ii). All the considerations are analogous with those in the proof of (i), because it suffices only to replace the curve γ_1 and the point $\gamma_2(0)$ by $\gamma_2 | [0,t]$ and $\gamma_1(0)$, respectively.

Verification of (iii). Let $[0,t) \subseteq P$, $[0,\tau) \subseteq R$ and let the equality $h = \tilde{\boldsymbol{f}}^T \tilde{\boldsymbol{g}}$ be valid on a neighbourhood $\tilde{U} \times \tilde{V}$ of the point $(\gamma_1(\tau), \gamma_2(t))$. Choose $\sigma < \tau$ and $s < t$ so that $\Gamma_1(\sigma,\tau) \subseteq \tilde{U}$ and $\Gamma_2(s,t) \subseteq \tilde{V}$. Since $\sigma \in R$ and $s \in P$, we have $h = \boldsymbol{f}^T \hat{\boldsymbol{g}}_\sigma$ on $\hat{U}_\sigma \times \hat{V}_\sigma$ and at the same time, $h = \boldsymbol{f}^T \boldsymbol{g}_s$ on $U_s \times V_s$, where $\hat{U}_\sigma$, $\hat{V}_\sigma$, U_s and V_s are sets described above in the definitions of the sets P and R. Note that the open sets $\tilde{U} \cap \hat{U}_\sigma \cap U_s$ and $\tilde{V} \cap \hat{V}_\sigma \cap V_s$ are nonempty, because they contain the points $\gamma_1(\sigma)$ and $\gamma_2(s)$, respectively. According to Theorem 2.3.1, there are nonsingular constant matrices C_1 and C_2 each of size $n \times n$ such that $\boldsymbol{f} = C_1^T \tilde{\boldsymbol{f}}$ on $\tilde{U} \cap \hat{U}_\sigma$, $\hat{\boldsymbol{g}}_\sigma = C_1^{-1} \tilde{\boldsymbol{g}}$ on $\tilde{V} \cap \hat{V}_\sigma$, $\boldsymbol{f} = C_2^T \tilde{\boldsymbol{f}}$ on $\tilde{U} \cap U_s$ and $\boldsymbol{g}_s = C_2^{-1} \tilde{\boldsymbol{g}}$ on $\tilde{V} \cap \tilde{V}_s$. It follows from the identity $C_2^T \tilde{\boldsymbol{f}} = C_1^T \tilde{\boldsymbol{f}}$ on the set $\tilde{U} \cap \hat{U}_\sigma \cap U_s$ that $C_1 = C_2$; hence we will write C instead of both matrices C_i. Now we introduce the sets $\hat{U}_\tau = (\hat{U}_\sigma \cup \tilde{U}) \cap U_s$ and $\hat{V}_\tau = (\hat{V}_\sigma \cap \tilde{V}) \cup V_s$. One can easily see that $\Gamma_1(\tau) \subseteq \hat{U}_\tau \subseteq U_0$ and $\Gamma_2(t) \subseteq \hat{V}_\tau$. Since $\hat{\boldsymbol{g}}_\sigma = \boldsymbol{g}_s (= C^{-1} \tilde{\boldsymbol{g}})$ on the set $\tilde{V} \cap \hat{V}_\sigma \cap V_s$, the following definition of a mapping

$\hat{\boldsymbol{g}}_\tau$ is correct:

$$\hat{\boldsymbol{g}}_\tau(y) := \begin{cases} \boldsymbol{g}_s(y) & \text{for } y \in V_s \\ \hat{\boldsymbol{g}}_\sigma(y) & \text{for } y \in \hat{V}_\sigma \cap \tilde{V}\,. \end{cases}$$

From this definition we can conclude that

$$\begin{aligned} &\boldsymbol{f}^T\hat{\boldsymbol{g}}_\tau = \boldsymbol{f}^T\boldsymbol{g}_s = h \quad \text{on } U_s \times V_s \\ &\boldsymbol{f}^T\hat{\boldsymbol{g}}_\tau = \boldsymbol{f}^T\hat{\boldsymbol{g}}_\sigma = h \quad \text{on } \hat{U}_\sigma \times (\hat{V}_\sigma \cap \tilde{V}) \\ &\boldsymbol{f}^T\hat{\boldsymbol{g}}_\tau = (C^T\tilde{\boldsymbol{f}})^T\hat{\boldsymbol{g}}_\sigma = (C^T\tilde{\boldsymbol{f}})^T(C^{-1}\tilde{\boldsymbol{g}}) = \tilde{\boldsymbol{f}}^T\tilde{\boldsymbol{g}} = h \quad \text{on } (\tilde{U} \cap U_s) \times (\hat{V}_\sigma \cap \tilde{V})\,. \end{aligned}$$

The last identity yields the representation $h = \boldsymbol{f}^T\hat{\boldsymbol{g}}_\tau$ on the set $\hat{U}_\tau \times \hat{V}_\tau$, because this set is a subset of the union of three sets

$$(U_s \times V_s) \cup [\hat{U}_\sigma \times (\hat{V}_\sigma \cap \tilde{V})] \cup [(\tilde{U} \cap U_s) \times (\hat{V}_\sigma \cap \tilde{V})]\,.$$

This means that $\tau \in R$, which completes the proof of (iii). □

Proof of Theorem 5.2.1: In view of Theorem 2.2.1 and part (i) of Remarks 2.2.2, the function h is of the form (5.2.1) on the set $X \times Y$ if there are elements $x_1, \dots, x_n \in X$ and $y_1, \dots, y_n \in Y$ such that the Casorati determinants of h (see (2.2.2)) satisfy

$$\det C_n h(x_1, \dots, x_n; y_1, \dots, y_n) \neq 0 \tag{5.2.3}$$

and

$$\det C_{n+1} h(x_1, \dots, x_n, x; y_1, \dots, y_n, y) = 0 \tag{5.2.4}$$

for each $x \in X$ and each $y \in Y$. Under the conditions of Theorem 5.2.1, we prove the existence of such elements $x_1, \dots, x_n$ and $y_1, \dots, y_n$ in the following way. The function h is assumed to be of the form (5.2.1) on $\tilde{U} \times \tilde{V}$, a neighbourhood of a (fixed) point $(x_0, y_0) \in X \times Y$, and the corresponding components f_i and g_i in (5.2.1) are linearly independent in the sets $\tilde{U}$ and $\tilde{V}$, respectively. As mentioned in the proof of Theorem 2.3.1 (see the sentence following formula (2.3.7)), the independence of f_i and g_i ensures the existence of elements $x_1, \dots, x_n \in \tilde{U}$ and $y_1, \dots, y_n \in \tilde{V}$ satisfying (5.2.3). Now for each (arbitrarily chosen) point $(x, y) \in X \times Y$, we construct two curves $\gamma_1 : [0, 1] \to X$ and $\gamma_2 : [0, 1] \to Y$ for their images Γ_1 and Γ_2 (see (5.2.2)) to contain the $(n+1)$-tuples of points $x_1, \dots, x_n, x$ and $y_1, \dots, y_n, y$, respectively. (Here we need the arcwise connectivity of the spaces X and Y.) Lemma 5.2.2 now implies that h is of the form (5.2.1) on the product $U \times V$, where U and V are some open neighbourhoods of the sets Γ_1 and Γ_2, respectively. In view of the first part of Theorem 2.2.1, this representation (5.2.1) on $U \times V$ implies the validity of (5.2.4), because $x_1, \dots, x_n, x \in U$ and $y_1, \dots, y_n, y \in V$. □

6 APPROXIMATE DECOMPOSITIONS OF SMOOTH FUNCTIONS

This chapter is concerned with the approximation problem

$$h(x,y) \approx \sum_{k=1}^{n} f_k(x) g_k(y) \tag{6a}$$

which seems to be of interest for each smooth function h not permitting any exact representation (2a). We start our procedure by introducing two natural approximating sums $T = T(h)$ and $S = S(h)$, analogous to Taylor series and the interpolation polynomials, respectively (see Section 6.1). As shown in Section 6.2, the corresponding errors $h - T$ and $h - S$ can be represented by formulas that resemble the Lagrange form for errors in polynomial approximations. The bounds for the errors $|h - T|$ and $|h - S|$ of different types are stated in Section 6.3. Some of these proofs will be postponed for Section 6.5, because they require finer estimates for some Wronski-like determinants, which are discussed in Section 6.4. The result proved in Section 6.5 will lead to the conclusion that the condition

$$\det W_n h(x,y) \neq 0 \quad \text{and} \quad \det W_{n+1} h(x,y) = 0 \quad \text{at each point } (x,y) \in I \times J \tag{6b}$$

of fundamental Theorem 2.1.1 concerning exact decompositions (2a) is "stable" in the following sense: If the Wronskians of the approximated function h satisfy

$$\det W_n h(x,y) \neq 0 \quad \text{and} \quad \left| \frac{\det W_{n+1} h(x,y)}{\det W_n h(x,y)} \right| \leq \varepsilon \quad \text{for each } (x,y) \in I \times J \tag{6c}$$

with some "small" constant ε, then the sup-norms (supremum norms) of the errors $h - T$ and $h - S$ are also of order ε. Let us finish this introductory part by remarking that while the problem of the best L^2-approximation (6a), with a prescribed number

n of products of arbitrary functions f_k and g_k, has been recently solved in [Ši 2] (see Chapter 7), the important problem of the best approximation (6a) with respect to the sup-norm seems to be still open.

6.1. Two types of approximating sums

In this section, we deal with the problem of creating suitable approximations (6a). As is usual in other occasions, we determine such approximations by imposing some *coincidence conditions*. In this way, we introduce two natural (and as it will be shown later, also effective) approximations T and S (see (6.1.2) and (6.1.7) below). This procedure has been recently described in [Ši 3].

We will assume that the number $n = 1, 2, \dots$ of terms in (6a) is a fixed integer, that $I = [a, b]$ and $J = [c, d]$ are two compact intervals in $\mathbb{R}$ and that the function $h : I \times J \to \mathbb{R}$ possesses the partial derivative $h_{x^n y^n}$, which is continuous on $I \times J$.

Approximations 6.1.1: **(i)** Suppose that $\det W_n h(x_0, y_0) \neq 0$ for some fixed $x_0 \in I$ and $y_0 \in J$. Let us show that the (unique) function $T : I \times J \to \mathbb{R}$ of the form (2a) that satisfies $2n$ functional conditions

$$T_{x^j}(x_0, \cdot\,) = h_{x^j}(x_0, \cdot\,) \quad \text{and} \quad T_{y^j}(\,\cdot\,, y_0) = h_{y^j}(\,\cdot\,, y_0) \quad (0 \le j \le n-1) \tag{6.1.1}$$

can be written as a matrix product

$$T(x,y) = \big(h(x,y_0), h_y(x,y_0), \dots, h_{y^{n-1}}(x,y_0)\big) \times W_n^{-1} h(x_0,y_0) \cdot \begin{pmatrix} h(x_0,y) \\ h_x(x_0,y) \\ \vdots \\ h_{x^{n-1}}(x_0,y) \end{pmatrix} \tag{6.1.2}$$

where $W_n^{-1} h$ denotes the inverse of the matrix $W_n h$. Indeed, if T is as in (2a) and satisfies (6.1.1), then

$$h_{x^j}(x_0, \cdot\,) = \sum_{k=1}^{n} f_k^{(j)}(x_0) g_k \quad \text{and} \quad h_{y^j}(\,\cdot\,, y_0) = \sum_{k=1}^{n} g_k^{(j)}(y_0) f_k \quad (0 \le j \le n-1). \tag{6.1.3}$$

We may consider (6.1.3) as two linear algebraic systems with unknown values of

$g_1, \ldots, g_n$ and $f_1, \ldots, f_n$, respectively. Notice that the matrices of these systems

$$F(x_0) = \begin{pmatrix} f_1(x_0) & f_2(x_0) & \ldots & f_n(x_0) \\ f_1'(x_0) & f_2'(x_0) & \ldots & f_n'(x_0) \\ \vdots & \vdots & \ddots & \vdots \\ f_1^{(n-1)}(x_0) & f_2^{(n-1)}(x_0) & \ldots & f_n^{(n-1)}(x_0) \end{pmatrix}$$

and

$$G(y_0) = \begin{pmatrix} g_1(y_0) & g_2(y_0) & \ldots & g_n(y_0) \\ g_1'(y_0) & g_2'(y_0) & \ldots & g_n'(y_0) \\ \vdots & \vdots & \ddots & \vdots \\ g_1^{(n-1)}(y_0) & g_2^{(n-1)}(y_0) & \ldots & g_n^{(n-1)}(y_0) \end{pmatrix}$$

are nonsingular, because (6.1.3) implies that

$$F(x_0) \cdot G^T(y_0) = W_n h(x_0, y_0) \tag{6.1.4}$$

and $W_n h(x_0, y_0)$ is supposed to be nonsingular. Hence (6.1.3) yields

$$\begin{pmatrix} g_1 \\ g_2 \\ \vdots \\ g_n \end{pmatrix} = F^{-1}(x_0) \cdot \begin{pmatrix} h(x_0, \cdot) \\ h_x(x_0, \cdot) \\ \vdots \\ h_{x^{n-1}}(x_0, \cdot) \end{pmatrix}$$

and

$$\begin{pmatrix} f_1 \\ f_2 \\ \vdots \\ f_n \end{pmatrix} = G^{-1}(y_0) \cdot \begin{pmatrix} h(\cdot, y_0) \\ h_y(\cdot, y_0) \\ \vdots \\ h_{y^{n-1}}(\cdot, y_0) \end{pmatrix}.$$

Substituting this into $T = \sum_{k=1}^{n} f_k g_k$ and taking into account (6.1.4), we conclude that (6.1.2) holds. On the other hand, it is easy to check that the function T defined by (6.1.2) satisfies (6.1.1).

(ii) Suppose that $x_1, x_2, \ldots, x_n \in I$ and $y_1, y_2, \ldots, y_n \in J$ are chosen so that the Casorati matrix

$$H := \begin{pmatrix} h(x_1, y_1) & h(x_1, y_2) & \ldots & h(x_1, y_n) \\ h(x_2, y_1) & h(x_2, y_2) & \ldots & h(x_2, y_n) \\ \vdots & \vdots & \ddots & \vdots \\ h(x_n, y_1) & h(x_n, y_2) & \ldots & h(x_n, y_n) \end{pmatrix} \tag{6.1.5}$$

(see (2.2.2)) is nonsingular. (The existence of such elements x_i and y_j is discussed in Remark 6.1.2 below.) Let us show that the (unique) function $S\colon I \times J \to \mathbb{R}$ of the form (2a) satisfying $2n$ functional conditions

$$S(x_j,\,\cdot\,) = h(x_j,\,\cdot\,) \quad \text{and} \quad S(\,\cdot\,,y_j) = h(\,\cdot\,,y_j) \quad (1 \le j \le n) \tag{6.1.6}$$

can be written as a matrix product

$$S(x,y) = \big(h(x,y_1), h(x,y_2), \ldots, h(x,y_n)\big) \cdot H^{-1} \cdot \begin{pmatrix} h(x_1,y) \\ h(x_2,y) \\ \vdots \\ h(x_n,y) \end{pmatrix}. \tag{6.1.7}$$

We can proceed analogously as in part (i). If S is as in (2a) and satisfies (6.1.6), then we can compute $g_1, g_2, \ldots, g_n$ and $f_1, f_2, \ldots, f_n$ from the systems

$$h(x_j,\,\cdot\,) = \sum_{k=1}^{n} f_k(x_j) g_k \quad \text{and} \quad h(\,\cdot\,,y_j) = \sum_{k=1}^{n} g_k(y_j) f_k \quad (1 \le j \le n)$$

and conclude that (6.1.7) holds. Conversely, the function S defined by (6.1.7) obviously satisfies (6.1.6).

Remark 6.1.2: By virtue of Theorem 2.2.1, the Casorati matrix H from (6.1.5) is singular for each $x_1, x_2, \ldots, x_n \in I$ and each $y_1, y_2, \ldots, y_n \in J$ just in the case when h is of the form

$$h(x,y) = \sum_{k=1}^{n'} f_k(x) g_k(y) \quad \text{for some } n' < n\,.$$

Consequently, for each function h not permitting any exact representation (2a), there exist elements $x_1, x_2, \ldots, x_n \in I$ and $y_1, y_2, \ldots, y_n \in J$ such that the matrix H from (6.1.5) is nonsingular and hence the sum S in (6.1.7) is well-defined.

6.2. Error representation

After introducing the approximating sums T and S in Section 6.1, we now turn our attention to the problem of representation of the errors $h - T$ and $h - S$.

Recall first the well-known Lagrange formulas

$$z(t_0) = z'(t_0) = \cdots = z^{(n-1)}(t_0) = 0 \quad \Longrightarrow \quad z(t) = \frac{z^{(n)}(\xi)}{n!} \cdot (t - t_0)^n \tag{6.2.1}$$

and

$$z(t_1) = z(t_2) = \cdots = z(t_n) = 0 \quad \Longrightarrow \quad z(t) = \frac{z^{(n)}(\xi)}{n!} \cdot \prod_{i=1}^{n} (t - t_i) \qquad (6.2.2)$$

where $t_0, t_1, \ldots, t_n, t, \xi \in I$ and $t_i \neq t_j$ $(1 \leq i < j \leq n)$, being valid for each function z possessing the n-th derivative on the interval I. (For general approximation properties of polynomials, see [MMR].) To state our Lagrange-like formulas for the errors $h - T$ and $h - S$ in approximation (6a), it is convenient to introduce the following "remainder" determinants

$$d_T(x,y) := \begin{vmatrix} h(x_0,y_0) & h_y(x_0,y_0) & \ldots & h_{y^{n-1}}(x_0,y_0) & h_{y^n}(x_0,y) \\ h_x(x_0,y_0) & h_{xy}(x_0,y_0) & \ldots & h_{xy^{n-1}}(x_0,y_0) & h_{xy^n}(x_0,y) \\ \vdots & \vdots & \ddots & \vdots & \vdots \\ h_{x^{n-1}}(x_0,y_0) & h_{x^{n-1}y}(x_0,y_0) & \ldots & h_{x^{n-1}y^{n-1}}(x_0,y_0) & h_{x^{n-1}y^n}(x_0,y) \\ h_{x^n}(x,y_0) & h_{x^n y}(x,y_0) & \ldots & h_{x^n y^{n-1}}(x,y_0) & h_{x^n y^n}(x,y) \end{vmatrix} \qquad (6.2.3)$$

and

$$d_S(x,y) := \begin{vmatrix} h(x_1,y_1) & h(x_1,y_2) & \ldots & h(x_1,y_n) & h_{y^n}(x_1,y) \\ h(x_2,y_1) & h(x_2,y_2) & \ldots & h(x_2,y_n) & h_{y^n}(x_2,y) \\ \vdots & \vdots & \ddots & \vdots & \vdots \\ h(x_n,y_1) & h(x_n,y_2) & \ldots & h(x_n,y_n) & h_{y^n}(x_n,y) \\ h_{x^n}(x,y_1) & h_{x^n}(x,y_2) & \ldots & h_{x^n}(x,y_n) & h_{x^n y^n}(x,y) \end{vmatrix}. \qquad (6.2.4)$$

Theorem 6.2.1: *Let I and J be two intervals in $\mathbb{R}$ and let $n \geq 1$ be an integer. Suppose that a function $h : I \times J \to \mathbb{K}$ has the partial derivative $h_{x^n y^n}$, which is continuous at each point of the rectangle $I \times J$. Assume also that the points $x_0 \in I$ and $y_0 \in J$ are chosen so that the Wronski matrix $W_n h(x_0, y_0)$ is nonsingular. Define T and d_T by* (6.1.2) *and* (6.2.3)*, respectively. Then for each $x \in I$ and $y \in J$, the difference $h(x,y) - T(x,y)$ can be represented as*

$$h(x,y) - T(x,y) = \frac{d_T(\xi,\eta)}{(n!)^2 \det W_n h(x_0,y_0)} \cdot (x - x_0)^n (y - y_0)^n \qquad (6.2.5)$$

where $\xi = \xi(x,y)$ lies between x_0 and x, while $\eta = \eta(x,y)$ lies between y_0 and y.

Proof: In view of (6.1.1), the function $\lambda := h - T$ satisfies

$$\lambda_{x^j}(x_0, \cdot\,) = 0 \quad \text{and} \quad \lambda_{y^j}(\,\cdot\,, y_0) = 0 \quad (0 \leq j \leq n-1). \qquad (6.2.6)$$

Thus we can apply (6.2.1) to the function $z = \lambda(\,\cdot\,, y)$ with a fixed $y \in J$ and conclude that

$$\lambda(x, y) = \frac{\lambda_{x^n}(\xi, y)}{n!} \cdot (x - x_0)^n \tag{6.2.7}$$

where $\xi = \xi(x, y)$ lies between x_0 and x. It follows from the second part of (6.2.6) that $\lambda_{x^n y^j}(\,\cdot\,, y_0) = 0$, for each $0 \le j \le n-1$. Applying now (6.2.1) to the function $z = \lambda_{x^n}(\xi, \,\cdot\,)$ with a fixed $\xi \in I$, we obtain

$$\lambda_{x^n}(\xi, y) = \frac{\lambda_{x^n y^n}(\xi, \eta)}{n!} \cdot (y - y_0)^n \tag{6.2.8}$$

where $\eta = \eta(\xi, y)$ lies between y_0 and y. Substituting (6.2.8) into (6.2.7), we observe that (6.2.5) holds if

$$\frac{d_T(\xi, \eta)}{\det W_n h(x_0, y_0)} = h_{x^n y^n}(\xi, \eta) - T_{x^n y^n}(\xi, \eta)\,. \tag{6.2.9}$$

However, the last equality follows from definitions (6.1.2) and (6.2.3) and from an elementary proposition of matrix theory:

If $A_n = [a_{ij}]_{i,j=1}^{n}$ and $A_{n+1} = [a_{ij}]_{i,j=1}^{n+1}$ and if $\det A_n \neq 0$, then

$$\frac{\det A_{n+1}}{\det A_n} = a_{n+1,n+1} - (a_{1,n+1}, \ldots, a_{n,n+1}) \cdot A_n^{-1} \cdot \begin{pmatrix} a_{n+1,1} \\ \vdots \\ a_{n+1,n} \end{pmatrix}. \tag{6.2.10}$$

Thus the proof is complete. □

Corollary 6.2.2: *Under the conditions of Theorem 6.2.1, the approximating sum T satisfies the following asymptotic formula*

$$\lim_{\substack{x \to x_0 \\ y \to y_0}} \frac{h(x, y) - T(x, y)}{(x - x_0)^n (y - y_0)^n} = \frac{\det W_{n+1}(x_0, y_0)}{(n!)^2 \cdot \det W_n h(x_0, y_0)}\,. \tag{6.2.11}$$

Proof: If $x \to x_0$ and $y \to y_0$ in (6.2.5), then $\xi \to x_0$, $\eta \to y_0$ and therefore by continuity, $d_T(\xi, \eta) \to d_T(x_0.y_0) = \det W_{n+1}(x_0, y_0)$, which proves (6.2.11). □

Theorem 6.2.3: *Let I and J be two intervals in $\mathbb{R}$ and let $n \ge 1$ be an integer. Suppose that a function $h : I \times J \to \mathbb{K}$ has the partial derivative $h_{x^n y^n}$, which is continuous at each point of the rectangle $I \times J$. Assume also that the points $x_1, x_2, \ldots, x_n \in I$ and $y_1, y_2, \ldots, y_n \in J$ are chosen so that the matrix H in*

(6.1.5) is nonsingular. Define S and d_S by (6.1.7) and (6.2.4), respectively. Then for each $x \in I$ and $y \in J$, the difference $h(x,y) - S(x,y)$ can be represented as

$$h(x,y) - S(x,y) = \frac{d_S(\xi,\eta)}{(n!)^2 \det H} \cdot \prod_{k=1}^{n} (x - x_k)(y - y_k) \tag{6.2.12}$$

where $\xi = \xi(x,y)$ and $\eta = \eta(x,y)$ lie in the minimal subintervals of I and J containing all the points $x, x_1, \ldots, x_n$ and $y, y_1, \ldots, y_n$, respectively.

Proof: In view of (6.1.6), the function $\mu := h - S$ satisfies

$$\mu(x_j, \cdot\,) = 0 \quad \text{and} \quad \mu(\,\cdot\,, y_j) = 0 \quad (1 \leq j \leq n)\,. \tag{6.2.13}$$

Thus we can apply (6.2.2) to the function $z = \mu(\,\cdot\,, y)$ with a fixed $y \in J$ and conclude that

$$\mu(x,y) = \frac{\mu_{x^n}(\xi, y)}{n!} \cdot \prod_{k=1}^{n} (x - x_k)$$

where $\xi = \xi(x,y)$ lies in the subinterval of I mentioned above. It follows from the second part of (6.2.13) that $\mu_{x^n}(\,\cdot\,, y_j) = 0$, for each $1 \leq j \leq n$. Applying now (6.2.2) to the function $z = \mu_{x^n}(\xi, \cdot\,)$ with a fixed $\xi \in I$, we obtain

$$\mu_{x^n}(\xi, y) = \frac{\mu_{x^n y^n}(\xi, \eta)}{n!} \cdot \prod_{k=1}^{n} (y - y_k)$$

where $\eta = \eta(\xi, y)$ lies in the subinterval of J mentioned above. Consequently, (6.2.12) holds if

$$\frac{d_S(\xi,\eta)}{\det H} = h_{x^n y^n}(\xi,\eta) - S_{x^n y^n}(\xi,\eta)\,. \tag{6.2.14}$$

However, the last equality follows from definitions (6.1.7) and (6.2.4) and from the rule (6.2.10). The proof is complete. □

Remark 6.2.4: The reader might feel the lack of some convergence formula like (6.2.11) which could illustrate the asymptotic behaviour of the error $h - S$. This is due to the fact that we cannot arrange the behaviour of x and y so that the "unknowns" ξ and η in (6.2.12) may converge to some definite limits. Nevertheless, one can propose, for example, the problem of determining n^2 limits

$$\lim_{\substack{x \to x_p \\ y \to y_q}} \frac{h(x,y) - S(x,y)}{(x - x_p)(y - y_q)} \qquad (p, q \in \{1, 2, \ldots, n\})\,.$$

We doubt whether any reasonable formulas for these limits exist at all.

Let us finish this section by proving a new criterion for decompositions (2a).

Theorem 6.2.5: *Let I and J be two intervals in $\mathbb{R}$ and let $n \geq 1$ be an integer. Suppose that a function $h : I \times J \to \mathbb{K}$ has the partial derivative $h_{x^n y^n}$, which is continuous at each point of the rectangle $I \times J$. Assume also that the points $x_0 \in I$ and $y_0 \in J$ are chosen so that the Wronski matrix $W_n h(x_0, y_0)$ is nonsingular. Then the function h is of the form* (2a), *with some components $f_k \in C^n(I)$ and $g_k \in C^n(J)$, if and only if the determinant d_T from* (6.2.3) *satisfies $d_T = 0$ on $I \times J$.*

Proof: If h is as in (2a), then the $(n+1) \times (n+1)$ matrix from (6.2.3) can be written as the product of the $(n+1) \times n$ matrix

$$\begin{pmatrix} f_1(x_0) & f_2(x_0) & \dots & f_n(x_0) \\ f_1'(x_0) & f_2'(x_0) & \dots & f_n'(x_0) \\ \vdots & \vdots & \ddots & \vdots \\ f_1^{(n-1)}(x_0) & f_2^{(n-1)}(x_0) & \dots & f_n^{(n-1)}(x_0) \\ f_1^{(n)}(x) & f_2^{(n)}(x) & \dots & f_n^{(n)}(x) \end{pmatrix}$$

times the $n \times (n+1)$ matrix

$$\begin{pmatrix} g_1(y_0) & g_1'(y_0) & \dots & g_1^{(n-1)}(y_0) & g_1^{(n)}(y) \\ g_2(y_0) & g_2'(y_0) & \dots & g_2^{(n-1)}(y_0) & g_2^{(n)}(y) \\ \vdots & \vdots & \ddots & \vdots & \vdots \\ g_n(y_0) & g_n'(y_0) & \dots & g_n^{(n-1)}(y_0) & g_n^{(n)}(y) \end{pmatrix}.$$

Since the rank of such a product does not exceed n, the common size of both the factors, we have $d_T = 0$ on $I \times J$. Conversely, if $d_T = 0$ on $I \times J$, then Theorem 6.2.1 implies that $h = T$ on $I \times J$. Since T is of type (2a), the proof is complete. □

Remark 6.2.6: Using Theorem 6.2.3, one can clearly obtain another new criterion for decompositions (2a) in the form $d_S = 0$ on $I \times J$. However, this result seems to have no essential advantage in comparison with that of part (i) of Remarks 2.2.2:

If the matrix H in (6.1.5) *is nonsingular, then h is of the form* (2a) *if and only if the determinant*

$$\det \begin{pmatrix} h(x_1, y_1) & h(x_1, y_2) & \dots & h(x_1, y_n) & h(x_1, y) \\ h(x_2, y_1) & h(x_2, y_2) & \dots & h(x_2, y_n) & h(x_2, y) \\ \vdots & \vdots & \ddots & \vdots & \vdots \\ h(x_n, y_1) & h(x_n, y_2) & \dots & h(x_n, y_n) & h(x_n, y) \\ h(x, y_1) & h(x, y_2) & \dots & h(x, y_n) & h(x, y) \end{pmatrix}$$

vanishes for each $(x, y) \in I \times J$.

(Compare the last determinant with that from (6.2.4).) Moreover, this result of Chapter 2 does not even require any smoothness restriction on the function h.

6.3. Error estimation

It is clear from the coincidence conditions (6.1.1) and (6.1.6) that the most effective estimates of the errors $|h - T|$ and $|h - S|$ we may expect should be of the form

$$|h(x,y) - T(x,y)| \leq \alpha_T(x,y)|(x-x_0)(y-y_0)|^n \quad (x \in I,\ y \in J) \qquad (6.3.1)$$

and

$$|h(x,y) - S(x,y)| \leq \alpha_S(x,y) \prod_{k=1}^{n} |(x-x_k)(y-y_k)| \quad (x \in I,\ y \in J) \quad (6.3.2)$$

where α_T and α_S are some bounded functions (or even constants). In fact, the Lagrange-like formulas (6.2.5) and (6.2.12) ensure that (6.3.1) and (6.3.2) are valid with

$$\alpha_T(x,y) = \frac{\sup\{|d_T(\xi,\eta)| : \xi \in I(x_0,x),\ \eta \in J(y_0,y)\}}{(n!)^2 \cdot |\det W_n h(x_0,y_0)|} \qquad (6.3.3)$$

and

$$\alpha_S(x,y) = \frac{\sup\{|d_S(\xi,\eta)| : \xi \in I(x_1,\ldots,x_n,x),\ \eta \in J(y_1,\ldots,y_n,y)\}}{(n!)^2 \cdot |\det H|} \qquad (6.3.4)$$

respectively, where $I(x_1,\ldots,x_n,x)$ and $J(y_1,\ldots,y_n,y)$ denote the subintervals mentioned in the statement of Theorem 6.2.3. Let us emphasize that the estimates (6.3.1) and (6.3.2) with the factors (6.3.3) and (6.3.4) are applicable whenever the sums T and S in (6.1.2) and (6.1.7) are well-defined (without any supplementary restriction like (6c) on the approximated function h). Moreover, for the sake of numerical applications, the sup-norms in (6.3.3) and (6.3.4) can be easily estimated by using the sup-norms of all the functions that occur in the determinants (6.2.3) and (6.2.4); see Corollary 6.3.1 below.

In what follows, we use the symbol $\|\cdot\|$ to denote the sup-norm of various matrices, scalar- and matrix-valued, one- and two-place functions. To avoid any confusion, we list now a few examples:

$$\|A\| = \max\{|a_{ij}| : 1 \leq i,j \leq n\} \quad \text{if } A \text{ is a constant matrix} \quad [a_{ij}]_{i,j=1}^{n},$$
$$\|h(\cdot,\cdot)\| = \sup\{|h(x,y)| : x \in I \text{ and } y \in J\},$$
$$\|h(\cdot,y)\| = \sup\{|h(x,y)| : x \in I\},$$
$$\|A(x,\cdot)\| = \sup\{\|A(x,y)\| : y \in J\},$$

etc.

Corollary 6.3.1: *Let I and J be two intervals in $\mathbb{R}$ and let $n \geq 1$ be an integer. Suppose that a function $h: I \times J \to \mathbb{K}$ has the partial derivative $h_{x^n y^n}$, which is continuous at each point of the rectangle $I \times J$.*

(i) *Let $x_0 \in I$ and $y_0 \in J$ be chosen so that the matrix $W_n h(x_0, y_0)$ is nonsingular. Define T by* (6.1.2). *Then the estimate* (6.3.1) *holds with a constant factor α_T given by*

$$\alpha_T = \frac{\|h_{x^n y^n}(\cdot,\cdot)\| + \|W_n^{-1} h(x_0, y_0)\| \sum\limits_{i=0}^{n-1} \|h_{x^n y^i}(\cdot, y_0)\| \sum\limits_{j=0}^{n-1} \|h_{x^j y^n}(x_0, \cdot)\|}{(n!)^2}. \tag{6.3.5}$$

(ii) *Let $x_1, x_2, \ldots, x_n \in I$ and $y_1, y_2, \ldots, y_n \in J$ be chosen so that the matrix H in* (6.1.5) *is nonsingular. Define S by* (6.1.7). *Then the estimate* (6.3.2) *holds with a constant factor α_S given by*

$$\alpha_S = \frac{\|h_{x^n y^n}(\cdot\,,\,\cdot)\| + \|H^{-1}\| \sum\limits_{i=1}^{n} \|h_{x^n}(\,\cdot\,, y_i)\| \sum\limits_{j=1}^{n} \|h_{y^n}(x_j,\,\cdot\,)\|}{(n!)^2}. \tag{6.3.6}$$

Proof: Part (i) immediately follows from formulas (6.2.5) and (6.2.9), where $T_{x^n y^n}$ is computed from (6.1.2). Similarly, part (ii) follows from (6.2.12) and (6.2.14), where $S_{x^n y^n}$ is computed from (6.1.7). The proof is complete. □

Although the bounds (6.3.5) and (6.3.6) are available for numerical calculations, their disadvantage (with respect to the theory of Wronski matrices given in Chapter 2) is evident: *The values of* (6.3.5) *and* (6.3.6) *are nonzero in general even if the function h satisfies condition* (6b) *of fundamental Theorem 2.1.1* (then, of course, $h = T = S$ on $I \times J$). This circumstance leads to the following question (in our opinion, an important one): is it possible to derive any other estimates of type (6.3.1) and (6.3.2), more "responsive" to the condition (6b)? Starting from the representation formulas (6.2.5), (6.2.11) and (6.2.12), it seems to be natural to seek some bounds for d_T and d_S depending on the sup-norm of the ratio $\det W_{n+1} / \det W_n h$. Before we state the main results in this direction, let us emphasize that we need an essential restriction on the function h. Namely, we will assume that the matrix $W_n h$ is nonsingular *at each point* of the rectangle $I \times J$ (see the first part of condition (6c), identical with the first part of condition (6b)).

Theorem 6.3.2: *Let I and J be two intervals in $\mathbb{R}$ and let $n \geq 1$ be an integer. Suppose that a function $h: I \times J \to \mathbb{K}$ has the partial derivative $h_{x^n y^n}$, which is continuous at each point of the rectangle $I \times J$, and that h satisfies* (6c) *for some constant $\varepsilon > 0$. Choose $x_0 \in I$, $y_0 \in J$ and define T by* (6.1.2). *Then the estimate*

(6.3.1) *holds with*

$$\alpha_T(x,y) = \frac{\varepsilon}{(n!)^2} \cdot \big(1 + K_1|x - x_0|\big)\big(1 + K_2|y - y_0|\big) \tag{6.3.7}$$

where the constants K_1, K_2 *are defined by*

$$\begin{aligned} K_1 &= \|W_n^{-1}h(\,\cdot\,,\,\cdot\,)\| \sum_{i=0}^{n-1} \|h_{x^n y^i}(\,\cdot\,,\,\cdot\,)\| \\ K_2 &= \|W_n^{-1}h(\,\cdot\,,\,\cdot\,)\| \sum_{j=0}^{n-1} \|h_{x^j y^n}(\,\cdot\,,\,\cdot\,)\| \,. \end{aligned} \tag{6.3.8}$$

Proof: see Section 6.5.

To prove the expected approximation property of the sum (6.1.7), we need (besides (6c)) another supplementary assumption imposed on the Wronski-like matrices

$$W_{x_1,\ldots,x_n}(y) := \begin{pmatrix} h(x_1,y) & h_y(x_1,y) & \ldots & h_{y^{n-1}}(x_1,y) \\ h(x_2,y) & h_y(x_2,y) & \ldots & h_{y^{n-1}}(x_2,y) \\ \vdots & \vdots & \ddots & \vdots \\ h(x_n,y) & h_y(x_n,y) & \ldots & h_{y^{n-1}}(x_n,y) \end{pmatrix}$$

and

$$W_{y_1,\ldots,y_n}(x) := \begin{pmatrix} h(x,y_1) & h(x,y_2) & \ldots & h(x,y_n) \\ h_x(x,y_1) & h_x(x,y_2) & \ldots & h_x(x,y_n) \\ \vdots & \vdots & \ddots & \vdots \\ h_{x^{n-1}}(x,y_1) & h_{x^{n-1}}(x,y_2) & \ldots & h_{x^{n-1}}(x,y_n) \end{pmatrix}.$$

Namely, we will assume that at least one of these matrices is nonsingular at each point of the corresponding interval I or J, respectively. It is clearly no loss of generality to assume that

$$\det W_{x_1,\ldots,x_n}(y) \neq 0 \quad \text{for each } y \in J\,. \tag{6.3.9}$$

Theorem 6.3.3: *Let* I *and* J *be two intervals in* $\mathbb{R}$ *and let* $n \geq 1$ *be an integer. Suppose that a function* $h\colon I \times J \to \mathbb{K}$ *has the partial derivative* $h_{x^n y^n}$*, which is continuous at each point of the rectangle* $I \times J$*, and that* h *satisfies* (6c) *for some constant* $\varepsilon > 0$*. Suppose also that* $x_1, x_2, \ldots, x_n \in I$ *and* $y_1, y_2, \ldots, y_n \in J$ *are chosen so that the matrix* H *in* (6.1.5) *is nonsingular and that* (6.3.9) *holds. Denote by* $\ell_1(x)$ *and* $\ell_2(y)$ *the lengths of the minimal subintervals containing all the points*

$x, x_1, x_2, \ldots, x_n$ and $y, y_1, y_2, \ldots, y_n$, respectively. Define S by (6.1.7). *Then the estimate* (6.3.2) *holds with*

$$\alpha_S(x,y) = \frac{\varepsilon}{(n!)^2} \cdot \big(1 + K_3\ell_1(x)\big)\big(1 + K_4\ell_2(y)\big) \tag{6.3.10}$$

where the constants K_3, K_4 are defined by

$$\begin{aligned} K_3 &= \|W_n^{-1}h(\,\cdot\,,\,\cdot\,)\|\big(1 + n^2\|W_{x_1,\ldots,x_n}^{-1}(\,\cdot\,)\| \cdot \|W_n h(\,\cdot\,,\,\cdot\,)\|\big) \\ &\quad\times \sum_{i=0}^{n-1} \|h_{x^n y^i}(\,\cdot\,,\,\cdot\,)\| \\ K_4 &= \|W_{x_1,\ldots,x_n}^{-1}(\,\cdot\,)\|\big(1 + n^2\|H^{-1}\| \cdot \|W_{x_1,\ldots,x_n}(-)\|\big) \\ &\quad\times \sum_{j=1}^{n} \|h_{y^n}(x_j,\,\cdot\,)\|\,. \end{aligned} \tag{6.3.11}$$

Proof: see Section 6.5.

Remark 6.3.4: In the statement of Theorem 6.3.3, asymmetry of the constants in (6.3.11) is due to the asymmetric condition (6.3.9).

6.4. Bounds for functional determinants

In order to make the main idea of our proofs of Theorems 6.3.2 and 6.3.3 more readable, we now separately discuss some required estimates for determinants that depend on a system of functions in one variable.

Throughout this section, let $z_1, \ldots, z_n \in C^n(I)$ be a fixed n-tuple of scalar functions such that their Wronski matrix

$$W(t) := \begin{pmatrix} z_1(t) & z_2(t) & \ldots & z_n(t) \\ z_1'(t) & z_2'(t) & \ldots & z_n'(t) \\ \vdots & \vdots & \ddots & \vdots \\ z_1^{(n-1)}(t) & z_2^{(n-1)}(t) & \ldots & z_n^{(n-1)}(t) \end{pmatrix}$$

is nonsingular at each point t of the interval $I = [a, b]$. Suppose also that the points $t_0, t_1, \ldots, t_n \in I$ are fixed and that the (constant) matrix

$$Z := \begin{pmatrix} z_1(t_1) & z_2(t_1) & \ldots & z_n(t_1) \\ z_1(t_2) & z_2(t_2) & \ldots & z_n(t_2) \\ \vdots & \vdots & \ddots & \vdots \\ z_1(t_n) & z_2(t_n) & \ldots & z_n(t_n) \end{pmatrix}$$

is nonsingular. Put $w(t) := \det W(t)$ and define

$$\varphi(t) := \frac{1}{w(t)} \cdot \det \begin{pmatrix} z_1(t) & z_2(t) & \dots & z_n(t) & z(t) \\ z_1'(t) & z_2'(t) & \dots & z_n'(t) & z'(t) \\ \vdots & \vdots & \ddots & \vdots & \vdots \\ z_1^{(n)}(t) & z_2^{(n)}(t) & \dots & z_n^{(n)}(t) & z^{(n)}(t) \end{pmatrix} \tag{6.4.1}$$

$$\psi(t) := \frac{1}{w(t_0)} \times \det \begin{pmatrix} z_1(t_0) & z_2(t_0) & \dots & z_n(t_0) & z(t_0) \\ z_1'(t_0) & z_2'(t_0) & \dots & z_n'(t) & z(t_0) \\ \vdots & \vdots & \ddots & \vdots & \vdots \\ z_1^{(n-1)}(t_0) & z_2^{(n-1)}(t_0) & \dots & z_n^{(n-1)}(t_0) & z^{(n-1)}(t_0) \\ z_1^{(n)}(t) & z_2^{(n)}(t) & \dots & z_n^{(n)}(t) & z^{(n)}(t) \end{pmatrix} \tag{6.4.2}$$

and

$$\theta(t) := \frac{1}{\det Z} \cdot \det \begin{pmatrix} z_1(t_1) & z_2(t_1) & \dots & z_n(t_1) & z(t_1) \\ z_1(t_2) & z_2(t_2) & \dots & z_n(t_2) & z(t_2) \\ \vdots & \vdots & \ddots & \vdots & \vdots \\ z_1(t_n) & z_2(t_n) & \dots & z_n(t_n) & z(t_n) \\ z_1^{(n)}(t) & z_2^{(n)}(t) & \dots & z_n^{(n)}(t) & z^{(n)}(t) \end{pmatrix} \tag{6.4.3}$$

for a given function $z \in C^n(I)$. The aim of our considerations is to estimate the values of ψ and θ by means of the sup-norm of φ.

Note that (6.4.1) can be considered as a linear nonhomogeneous differential equation of order n with respect to the "unknown" z. The well-known method of variation of parameters shows that each solution z can be written in the form

$$z(t) = \big(z_1(t), \dots, z_n(t)\big) \cdot \left(\mathbf{c} + \int_{t_0}^{t} \mathbf{c}_n W^{-1}(s)\varphi(s)\, ds\right) \tag{6.4.4}$$

where $\mathbf{c}_n W^{-1}$ denotes the n-th column of the matrix W^{-1} and $\mathbf{c} \in \mathbb{R}^n$ is arbitrary. In view of a basic determinant property, it is easily seen from (6.4.2) and (6.4.3) that the functions ψ and θ do not depend on the choice of the vector $\mathbf{c}$ in (6.4.4). To calculate ψ, it is convenient to put $\mathbf{c} = \mathbf{0}$. In fact, the function

$$z_0(t) := \big(z_1(t), \dots, z_n(t)\big) \cdot \int_{t_0}^{t} \mathbf{c}_n W^{-1}(s)\varphi(s)\, ds \tag{6.4.5}$$

satisfies

$$z_0^{(j)}(t) = \left(z_1^{(j)}(t), \ldots, z_n^{(j)}(t)\right) \cdot \int_{t_0}^{t} \mathbf{c}_n W^{-1}(s)\varphi(s)\, ds \quad (1 \le j \le n-1)$$

and

$$z_0^{(n)}(t) = \varphi(t) + \left(z_1^{(n)}(t), \ldots, z_n^{(n)}(t)\right) \cdot \int_{t_0}^{t} \mathbf{c}_n W^{-1}(s)\varphi(s)\, ds\,. \tag{6.4.6}$$

Hence $z_0(t_0) = z_0'(t_0) = \ldots = z_0^{(n-1)}(t_0) = 0$. Substituting $z = z_0$ into (6.4.2), we therefore conclude that $\psi(t) = z_0^{(n)}(t)$, for each $t \in I$. Consequently, formula (6.4.6) immediately yields the following:

Lemma 6.4.1: *The inequality*

$$|\psi(t)| \le \|\varphi(-)\| \left(1 + \|W^{-1}(-)\| \sum_{k=1}^{n} |z_k^{(n)}(t)| \cdot |t - t_0|\right)$$

holds for each $t \in I$.

To estimate the function θ, we first utilize the rule (6.2.10) and observe that

$$\theta(t) = z^{(n)}(t) - \left(z_1^{(n)}(t), \ldots, z_n^{(n)}(t)\right) \cdot Z^{-1} \cdot \begin{pmatrix} z(t_1) \\ z(t_2) \\ \vdots \\ z(t_n) \end{pmatrix}. \tag{6.4.7}$$

Let $z = z_0$ again, where z_0 is defined by (6.4.5) with t_0 replaced by any t_j, say $t_0 = t_1$. In view of (6.4.5) and (6.4.6), we have

$$|z_0(t_j)| \le n\|\varphi(-)\| \cdot \|W^{-1}(-)\| \cdot \|W(-)\| \cdot |t_j - t_1| \quad (j = 1, \ldots, n)$$

and

$$|z_0^{(n)}(t)| \le \|\varphi(-)\| \left(1 + \|W^{-1}(-)\| \sum_{k=1}^{n} |z_k^{(n)}(t)| \cdot |t - t_1|\right) \quad (t \in I)\,.$$

Consequently, formula (6.4.7) with $z = z_0$ yields the following:

Lemma 6.4.2: *Denote by $\ell(t)$ the length of the minimal subinterval containing all the points $t, t_1, t_2, \ldots, t_n$. The inequality*

$$|\theta(t)| \leq \|\varphi(-)\| \left(1 + \|W^{-1}(-)\| (1 + n^2 \|Z^{-1}\| \cdot \|W(-)\|) \sum_{k=1}^{n} |z_k^{(n)}(t)| \, \ell(t)\right)$$

holds for each $t \in I$.

6.5. Proofs of approximation theorems

Now we are ready to prove both the approximation theorems stated in the last part of Section 6.3.

Proof of Theorem 6.3.2: To find a relationship between the Wronskian $\det W_{n+1}$ and the determinant d_T from (6.2.3), we introduce an "intermediate" determinant

$$\tilde{d}(x,y) := \begin{vmatrix} h(x_0,y) & h_y(x_0,y) & \cdots & h_{y^{n-1}}(x_0,y) & h_{y^n}(x_0,y) \\ h_x(x_0,y) & h_{xy}(x_0,y) & \cdots & h_{xy^{n-1}}(x_0,y) & h_{xy^n}(x_0,y) \\ \vdots & \vdots & \ddots & \vdots & \vdots \\ h_{x^{n-1}}(x_0,y) & h_{x^{n-1}y}(x_0,y) & \cdots & h_{x^{n-1}y^{n-1}}(x_0,y) & h_{x^{n-1}y^n}(x_0,y) \\ h_{x^n}(x,y) & h_{x^n y}(x,y) & \cdots & h_{x^n y^{n-1}}(x,y) & h_{x^n y^n}(x,y) \end{vmatrix}.$$

For each fixed $y \in J$, Lemma 6.4.1 with $z_i = h_{y^{i-1}}(\cdot\,, y)$, $z = h_{y^n}(\cdot\,, y)$ leads to the conclusion that the estimate

$$\left| \frac{\tilde{d}(x,y)}{\det W_n h(x_0,y)} \right| \leq \left\| \frac{\det W_{n+1}(\cdot\,,y)}{\det W_n h(\cdot\,,y)} \right\| (1 + K_1 |x - x_0|) \tag{6.5.1}$$

holds for each $x \in I$ if K_1 is defined by (6.3.8). Now for each fixed $x \in I$, we can apply Lemma 6.4.1 again, in this case with $z_j = h_{x^{j-1}}(x_0, \cdot)$, $z = h_{x^n}(x, \cdot)$ and conclude that the estimate

$$\left| \frac{d_T(x,y)}{\det W_n h(x_0,y_0)} \right| \leq \left\| \frac{\tilde{d}(x, \cdot)}{\det W_n h(x_0, \cdot)} \right\| (1 + K_2 |y - y_0|) \tag{6.5.2}$$

holds for each $y \in J$ if K_2 is defined by (6.3.8). In view of ε-condition (6c), it follows from (6.5.1) and (6.5.2) that the inequality

$$\left| \frac{d_T(x,y)}{\det W_n h(x_0,y_0)} \right| \leq \varepsilon \big(1 + K_1 |x - x_0|\big) \big(1 + K_2 |y - y_0|\big)$$

holds at each point $(x, y) \in I \times J$. By Theorem 6.2.1, the last estimate yields (6.3.1), with α_T as indicated in (6.3.7). The proof is complete. □

Proof of Theorem 6.3.3: Let us introduce another "intermediate" determinant

$$\tilde{d}(x,y) := \begin{vmatrix} h(x_1,y) & h_y(x_1,y) & \dots & h_{y^{n-1}}(x_1,y) & h_{y^n}(x_1,y) \\ h(x_2,y) & h_y(x_2,y) & \dots & h_{y^{n-1}}(x_2,y) & h_{y^n}(x_2,y) \\ \vdots & \vdots & \ddots & \vdots & \vdots \\ h(x_n,y) & h_y(x_n,y) & \dots & h_{y^{n-1}}(x_n,y) & h_{y^n}(x_n,y) \\ h_{x^n}(x,y) & h_{x^n y}(x,y) & \dots & h_{x^n y^{n-1}}(x,y) & h_{x^n y^n}(x_n,y) \end{vmatrix}.$$

For each fixed $y \in J$, Lemma 6.4.2 with $z_i = h_{y^{i-1}}(\,\cdot\,, y)$, $z = h_{y^n}(\,\cdot\,, y)$ and $t_k = x_k$ leads to the conclusion that the estimate

$$\left| \frac{\tilde{d}(x,y)}{\det W_{x_1,\dots,x_n}(y)} \right| \le \left\| \frac{\det W_{n+1}(\,\cdot\,,y)}{\det W_n h(\,\cdot\,,y)} \right\| \big(1 + K_3 \ell_1(x)\big) \tag{6.5.3}$$

holds for each $x \in I$ if K_3 is as in (6.3.11). Now for each fixed $x \in I$, we can apply Lemma 6.4.1 again, in this case with $z_j = h(x_j, \,\cdot\,)$, $z = h(x, \,\cdot\,)$ and $t_k = y_k$ to conclude that the estimate

$$\left| \frac{d_S(x,y)}{\det H} \right| \le \left\| \frac{\tilde{d}(x, \,\cdot\,)}{\det W_{x_1,\dots,x_n}(-)} \right\| \big(1 + K_4 \ell_2(y)\big) \tag{6.5.4}$$

holds for each $y \in J$ if K_4 is as in (6.3.11). In view of ε-condition (6c), it follows from (6.5.3) and (6.5.4) that the inequality

$$\left| \frac{d_S(x,y)}{\det H} \right| \le \varepsilon \big(1 + K_3 \ell_1(x)\big) \big(1 + K_4 \ell_2(y)\big)$$

holds at each point $(x, y) \in I \times J$. By Theorem 6.2.3, the last estimate yields (6.3.2), with α_S as indicated in (6.3.10). The proof is complete. □

7 THE BEST L^2-APPROXIMATIONS OF TWO-PLACE FUNCTIONS

Functions of the form

$$\sum_{k=1}^{n} f_k(x) g_k(y) \tag{7a}$$

play an important role in the theory of integral equations. In fact, a linear integral operator $v = Tu$ defined by the rule

$$v(y) = \int_E h(x,y)u(x)\,d\mu(x) \quad (E \text{ is a fixed measure space}) \tag{7b}$$

has a finite-dimensional image space provided that its kernel function h is exactly equal to a sum (7a). Consequently, the spectral theory of such *degenerate* operators is extremely simple. This idea lies at the heart of the theory of compact integral operators (7b) whose kernel functions h are approximable by finite sums (7a) with arbitrary precision (see [Ta, pages 274–285]).

We restrict our attention to the functions of the class L^2. One can show that if $X \times Y$ is the Cartesian product of measure spaces X and Y, then the set of all sums (7a), in which $f_k \in L^2(X)$, $g_k \in L^2(Y)$ and $n \in \{1, 2, \dots\}$, is a dense subset on $L^2(X \times Y)$. Indeed, if $\{u_i \mid i \in I\}$ and $\{v_j \mid j \in J\}$ are orthonormal bases of the spaces $L^2(X)$ and $L^2(Y)$ respectively, then $\{u_i v_j \mid i \in I \text{ and } j \in J\}$ is an orthonormal basis of $L^2(X \times Y)$. In this chapter, we will provide a solution (see [Ši 2]) of the following problem concerning the best L^2–approximations: *Given a function $h \in L^2(X \times Y)$ and an integer $n \geq 1$, find two n-tuples of functions $f_1, f_2, \dots, f_n \in L^2(X)$ and $g_1, g_2, \dots, g_n \in L^2(Y)$, for which the value of*

$$\left\| h - \sum_{k=1}^{n} f_k g_k \right\|_{L^2(X\times Y)} = \sqrt{\int_{X\times Y} \left| h - \sum_{k=1}^{n} f_k g_k \right|^2} \tag{7c}$$

is as small as possible.

The existence of these best approximations will be established in Section 7.1 by using the technique of the Casorati determinants for some bilinear functional generated by the approximated function h. In Section 7.2 we will show that the components f_k and g_k of the best approximation $s_n = \sum_{k=1}^{n} f_k g_k$ have to satisfy a system of Fredholm integral equations. The elementary properties of the corresponding integral operators are discussed in Section 7.3 (see also [RSz.-N]). The solution of the main approximation problem is then given as Theorem 7.4.1. As a consequence, we obtain a proof of the so-called Hilbert–Schmidt decomposition for functions of the class $L^2(X \times Y)$ in a way that differs from the usual procedure known from the theory of compact operators. All the results of this chapter have been proved in [Ši 2].

Main Definitions: Troughout the chapter, (X, μ) and (Y, λ) are assumed to be two fixed measure spaces with σ-additive and σ-finite measures μ and λ, respectively. (We need σ-finiteness of μ and λ for the Fubini theorem

$$\int_{X\times Y} h\, d(\mu \times \lambda) = \int_X d\mu(x) \int_Y h(x,y)\, d\lambda(y) = \int_Y d\lambda(y) \int_X h(x,y)\, d\mu(x)$$

to be applicable; see [Ta, Section 7.8] or [Ru]). As mentioned in Section 1.3, $L^2(X)$ denotes the space of all real- or complex-valued measurable functions $f : X \to \mathbb{K}$ satisfying $\int_X |f|^2\, d\mu < \infty$, which is a Hilbert space with the inner product

$$\langle f_1, f_2 \rangle = \int_X f_1 \bar{f}_2\, d\mu\,.$$

Since we will consider the spaces $L^2(X)$, $L^2(Y)$ and $L^2(X \times Y)$ simultaneously, we sometimes use a subscript as in (7c) to denote the norm in a space H as $\|\ \|_H$.

Note that $f \cdot g \in L^2(X \times Y)$ whenever $f \in L^2(X)$ and $g \in L^2(Y)$. This is why the classes of functions $\mathscr{S}_1 \subseteq \mathscr{S}_2 \subseteq \dots$ defined by

$$\mathscr{S}_n = \left\{ \sum_{k=1}^{n} f_k g_k \;\middle|\; f_k \in L^2(X) \quad \text{and} \quad g_k \in L^2(Y), \quad 1 \le k \le n \right\} \tag{7d}$$

form a family of subsets in $L^2(X \times Y)$. For the sake of short formulations we will consider also one-element set $\mathscr{S}_0$ containing the zero (function) of $L^2(X \times Y)$.

It will be convenient to use standard vector and matrix notations. Each n-tuple of functions $f_1, f_2, \dots, f_n$ will be written as a vector (i.e. $n \times 1$ matrix) function $\boldsymbol{f}$ and the following conventions will be accepted:

$$\boldsymbol{f} \in L^2(X) \text{ if and only if } f_k \in L^2(X),\ 1 \le k \le n\,,$$
$$\boldsymbol{f}^* = (\,\bar{f}_1,\ \ \bar{f}_2,\ \ \ldots,\ \ \bar{f}_n\,) \quad \text{(conjugate transposition of } \boldsymbol{f}\text{)}\,,$$
$$\text{and} \quad T\boldsymbol{f} = \begin{pmatrix} Tf_1 \\ Tf_2 \\ \vdots \\ Tf_n \end{pmatrix} \quad \text{for any operator or functional } T\,.$$

Then each element of $\mathscr{S}_n$ can be written as a matrix product $\boldsymbol{f}^*(x)\boldsymbol{g}(y)$, where $\boldsymbol{f} \in L^2(X)$ and $\boldsymbol{g} \in L^2(Y)$ are n-dimensional vector functions. (It is not substantial that functions f_k from the sum in (7d) are replaced by $\bar{f}_k$.) Finally, $\Gamma(\boldsymbol{f})$ will denote the Gram matrix (1.3.1) of the components f_j of the vector function $\boldsymbol{f}$.

Given $h \in L^2(X \times Y)$, define the sequence of distances

$$\rho_n(h) := \inf\Big\{\big\|h - s_n\big\|_{L^2(X\times Y)} : s_n \in \mathscr{S}_n\Big\} \quad (n = 0, 1, 2, \ldots)\,. \tag{7e}$$

It is clear that

$$\big\|h\big\| = \rho_0(h) \ge \rho_1(h) \ge \rho_2(h) \ge \ldots$$

and that (as explained above) $\rho_n(h) \to 0$ as $n \to \infty$. It is worth remarking that $\rho_{n+1}(h) < \rho_n(h)$ for any $h \notin \mathscr{S}_n$ (see Theorem 7.1.1). In the case when X and Y are open intervals on the real line with the usual Lebesque measure, Neuman [N 3] has constructed a function $h = h_n \in L^2(X \times Y)$ satisfying $\rho_n(h) > 0$ for a given fixed n. We will show that even $\rho_n(h) > 0$ for each $h \in L^2(X \times Y) \setminus \mathscr{S}_n$. (Notice that the condition $\mathscr{S}_n \subsetneq L^2(X \times Y)$ fails to hold for sufficiently large n if the sets X and Y have finitely many elements.)

7.1. Existence theory for best approximations

The aim of this section is to establish the existence of the best approximations by functions of the classes $\mathscr{S}_n$. We state it as the following:

Theorem 7.1.1: *Let $h \in L^2(X \times Y)$ be a function and let $n \ge 1$ be an integer. Then there exists at least one function $s_n \in \mathscr{S}_n$ with the property*

$$\big\|h - s_n\big\| = \rho_n(h)\,. \tag{7.1.1}$$

If in addition $h \notin \mathscr{S}_{n-1}$, then each function $s_n \in \mathscr{S}_n$ satisfying (7.1.1) *does not lie in $\mathscr{S}_{n-1}$, which means that $\rho_{n-1}(h) > \rho_n(h)$.*

To prove Theorem 7.1.1, we need essentially to show that $\mathscr{S}_n$ is a weakly closed subset of $L^2(X \times Y)$. (We call the reader's attention to the fact that $\mathscr{S}_n$ is neither a subspace nor a convex subset of $L^2(X \times Y)$.) To this purpose, we apply Theorem 2.2.1, the fundamental result of Chapter 2 on the Casorati determinants, to the following (bilinear) mapping

$$B(u,v) = \int\limits_{X\times Y} hu\bar{v}\, d(\mu \times \lambda) \qquad (u \in L^2(X) \text{ and } v \in L^2(Y)) . \tag{7.1.2}$$

This application yields the following:

Proposition 7.1.2: *Let $h \in L^2(X \times Y)$ be any function. Define a mapping $B : L^2(X) \times L^2(Y) \to \mathbb{K}$ by the rule* (7.1.2). *Then h lies in $\mathscr{S}_n$ for some given n if and only if the equality*

$$\det \begin{pmatrix} B(u_1,v_1) & B(u_1,v_2) & \dots & B(u_1,v_{n+1}) \\ B(u_2,v_1) & B(u_2,v_2) & \dots & B(u_2,v_{n+1}) \\ \vdots & \vdots & \ddots & \vdots \\ B(u_{n+1},v_1) & B(u_{n+1},v_2) & \dots & B(u_{n+1},v_{n+1}) \end{pmatrix} = 0 \tag{7.1.3}$$

holds for any $u_1, u_2, \dots, u_{n+1} \in L^2(X)$ and any $v_1, v_2, \dots, v_{n+1} \in L^2(Y)$.

Proof: (i) Assume that $h \in \mathscr{S}_n$, i.e. $h = \sum_{k=1}^n f_k g_k$, where $f_k \in L^2(X)$ and $g_k \in L^2(Y)$, $1 \le k \le n$. Substituting this into (7.1.2) and using the Fubini theorem, we find that

$$B(u,v) = \sum_{k=1}^{n} F_k(u)G_k(v) \qquad (u \in L^2(X) \text{ and } v \in L^2(Y)) \tag{7.1.4}$$

where

$$F_k(u) = \int_X f_k u \, d\mu \quad \text{and} \quad G_k(v) = \int_Y g_k \bar{v}\, d\lambda .$$

Consequently, B is of the form (2a) and Theorem 2.2.1 implies that the equality (7.1.3) holds whenever $u_1, u_2, \dots, u_{n+1} \in L^2(X)$ and $v_1, v_2, \dots, v_{n+1} \in L^2(Y)$.

(ii) Assume that the bilinear form (7.1.2) satisfies (7.1.3) identically. We may also assume that the inequality

$$\det \begin{pmatrix} B(u_1,v_1) & B(u_1,v_2) & \dots & B(u_1,v_n) \\ B(u_2,v_1) & B(u_2,v_2) & \dots & B(u_2,v_n) \\ \vdots & \vdots & \ddots & \vdots \\ B(u_n,v_1) & B(u_n,v_2) & \dots & B(u_n,v_n) \end{pmatrix} \neq 0 \tag{7.1.5}$$

holds for some $u_1, u_2, \dots, u_n \in L^2(X)$ and $v_1, v_2, \dots, v_n \in L^2(Y)$; otherwise the number n in (7.1.3) can be reduced. Then Theorem 2.2.1 implies that B is of type

(7.1.4), with F_k and G_k of the following form

$$F_k(u) = \sum_{i=1}^{n} c_{ki} B(u, v_i) \quad \text{and} \quad G_k(v) = \sum_{i=1}^{n} d_{ki} B(u_i, v) \qquad (1 \le k \le n)$$

in which c_{ki} and d_{ki} are suitable constants. Now the Fubini theorem implies that

$$\left. \begin{aligned} F_k(u) &= \sum_{i=1}^{n} c_{ki} B(u, v_i) = \int_X f_k u \, d\mu \\ G_k(v) &= \sum_{i=1}^{n} d_{ki} B(u_i, v) = \int_Y g_k \bar{v} \, d\lambda \end{aligned} \right. \qquad (1 \le k \le n)$$

where $f_k \in L^2(X)$ and $g_k \in L^2(Y)$ are independent of u and v:

$$f_k(x) = \sum_{i=1}^{n} c_{ki} \int_Y h(x,y) \bar{v}_i(y) \, d\lambda(y) \qquad \text{for almost all } x \in X$$

$$g_k(y) = \sum_{i=1}^{n} d_{ki} \int_X h(x,y) u_i(x) \, d\mu(x) \qquad \text{for almost all } y \in Y.$$

Substituting this representation of F_k and G_k into (7.1.4), we obtain

$$\begin{aligned} \int_{X \times Y} h u \bar{v} \, d(\mu \times \lambda) &= \sum_{k=1}^{n} F_k(u) G_k(v) = \sum_{k=1}^{n} \int_X f_k u \, d\mu \cdot \int_Y g_k \bar{v} \, d\lambda \\ &= \int_{X \times Y} \left(\sum_{k=1}^{n} f_k g_k \right) u \bar{v} \, d(\mu \times \lambda) \end{aligned}$$

for any $u \in L^2(X)$ and $v \in L^2(Y)$. The last identity is possible only if $h = \sum_{k=1}^{n} f_k g_k$ almost everywhere on $X \times Y$, which means that $h \in \mathscr{S}_n$. □

Proposition 7.1.2 enables us to verify a topological property of the subsets $\mathscr{S}_n$:

Lemma 7.1.3: *The family of functions $\mathscr{S}_n$ is a weakly closed subset of the Hilbert space $L^2(X \times Y)$, for each $n \in \mathbb{N}$.*

Proof: Let $n \in \mathbb{N}$ be fixed and let $\{h_k\}_{k=1}^{\infty} \subseteq \mathscr{S}_n$ be any sequence that converges weakly to a function $h_0 \in L^2(X \times Y)$, i.e.

$$\langle h_k, h \rangle \to \langle h_0, h \rangle \text{ as } k \to \infty, \text{ for any } h \in L^2(X \times Y).$$

Particularly, we have

$$B_k(u,v) \to B_0(u,v) \text{ as } k \to \infty, \text{ where } B_k(u,v) = \int\limits_{X\times Y} h_k u\bar{v}\, d(\mu \times \lambda)$$

for any $u \in L^2(X)$ and any $v \in L^2(Y)$. Since $h_k \in \mathscr{S}_n$ for each $k \geq 1$, Proposition 7.1.2 implies that the equality (7.1.3) holds with fixed $u_1, u_2, \ldots, u_{n+1} \in L^2(X)$ and $v_1, v_2, \ldots, v_{n+1} \in L^2(Y)$, for each $B = B_k$, $k \geq 1$. Now we may let $k \to \infty$ in (7.1.3) with $B = B_k$ to conclude that (7.1.3) is valid with $B = B_0$ as well. Since the elements $u_1, u_2, \ldots, u_{n+1} \in L^2(X)$ and $v_1, v_2, \ldots, v_{n+1} \in L^2(Y)$ in (7.1.3) were arbitrary, Proposition 7.1.2 implies that $h_0 \in \mathscr{S}_n$. □

Now we are in a position to prove the above-stated Theorem 7.1.1.

Proof of Theorem 7.1.1: Since the subsets $\mathscr{S}_n$ are weakly closed, we can proceed in a standard way. Having fixed $h \in L^2(X \times Y)$ and $n \in \mathbb{N}$, we first choose a sequence $\{s_n^k\}_{k=1}^{\infty} \subseteq \mathscr{S}_n$ so that

$$\|h - s_n^k\| \to \rho_n(h) \text{ as } k \to \infty\,. \tag{7.1.6}$$

This sequence is clearly bounded in $L^2(X \times Y)$, because of the estimates $\|s_n^k\| \leq \|h\| + \|s_n^k - h\|$, $k \geq 1$. Since any bounded subset of a Hilbert space is weakly precompact, there exists a subsequence $\{s_n^{k_j}\}_{j=1}^{\infty}$ that converges weakly to a function $s_n^0 \in L^2(X \times Y)$. Lemma 7.1.3 now implies that $s_n^0 \in \mathscr{S}_n$. The definition (7e) of $\rho_n(h)$ and the convergence (7.1.6) yield

$$\|h - s_n^0\| \geq \rho_n(h) = \lim_{j\to\infty} \|h - s_n^{k_j}\|\,.$$

On the other hand, the function $h - s_n^0$ is clearly the weak limit of the sequence $\{h - s_n^{k_j}\}_{j=1}^{\infty}$, which provides a well-known inequality

$$\|h - s_n^0\| \leq \liminf_{j\to\infty} \|h - s_n^{k_j}\|\,.$$

Thus $\|h - s_n^0\| = \rho_n(h)$ and the first part of Theorem 7.1.1 is proved. It remains to verify the implications

$$\rho_{n-1}(h) = \rho_n(h) \;\Rightarrow\; h \in \mathscr{S}_{n-1}\,. \tag{7.1.7$_n$}$$

We start with the case $n = 1$. So let $h \in L^2(X \times Y)$ be a function satisfying $\rho_1(h) = \|h\| (= \rho_0(h))$. If $u \in L^2(X)$ and $v \in L^2(Y)$, then $cuv \in \mathscr{S}_1$ for any constant $c \in \mathbb{K}$. Consequently, we have

$$\|h - cuv\| \geq \rho_1(h) = \|h\| \quad (\text{for each } c \in \mathbb{K}). \tag{7.1.8}$$

The known orthoprojection principle (cf. Section 7.2 below) implies that (7.1.8) is possible if and only if h and uv are orthogonal elements of $L^2(X \times Y)$. Since $u \in L^2(X)$ and $v \in L^2(Y)$ are arbitrary, this means that h is the zero-function, which proves (7.1.7$_1$). To prove (7.1.7$_n$) with $n > 1$, suppose that $\rho_{n-1}(h) = \rho_n(h)$, i.e. that $\|h - s_{n-1}\| = \rho_n(h)$ for some $s_{n-1} \in \mathscr{S}_{n-1}$. Denote $\tilde{h} = h - s_{n-1}$. Then for each $s_1 \in \mathscr{S}_1$ we have

$$\|\tilde{h} - s_1\| = \|h - (s_{n-1} + s_1)\| \geq \rho_n(h)$$

because of $s_{n-1} + s_1 \in \mathscr{S}_n$. Therefore, $\rho_1(\tilde{h}) \geq \rho_n(h) = \|\tilde{h}\|$. Since the converse inequality $\rho_1(\tilde{h}) \leq \|\tilde{h}\|$ is trivial, we find that $\rho_1(\tilde{h}) = \|\tilde{h}\|$. However, we have already shown that this equality holds only if $\tilde{h}$ is the zero-function, which means that $h = s_{n-1} \in \mathscr{S}_{n-1}$. □

7.2. Integral equations for best approximations

In this section we find necessary conditions for a function $s_n \in \mathscr{S}_n$ to satisfy $\|h - s_n\| = \rho_n(h)$. It will be shown (see Theorem 7.2.2 below) that the functions f_k and g_k forming such an $s_n = \sum_{k=1}^{n} f_k g_k$ have to obey a system of linear integral equations determined by the function h.

Let us start by recalling a well-known principle underlying the theory of Fourier series.

Orthoprojection Principle 7.2.1: *Let H be a Hilbert space and let L be a linear subspace of H generated by the given elements $u_1, u_2, \ldots, u_n \in H$. Suppose also that $h \in H$ is fixed. Then an element $h_0 \in L$ has the property*

$$\|h - h_0\| = \inf\left\{\|h - u\| \,:\, u \in L\right\}$$

if and only if

$$\langle h - h_0, u_j\rangle = 0 \qquad (j = 1, 2, \ldots, n). \tag{7.2.1}$$

Note that if $h_0 = c_1 u_1 + \cdots + c_n u_n$, then (7.2.1) presents a system of n linear equations

$$\sum_{i=1}^{n} \langle u_i, u_j\rangle c_i = \langle h, u_j\rangle \quad (j = 1, 2, \ldots, n) \tag{7.2.2}$$

with unknown constants $c_1, \ldots, c_n \in \mathbb{K}$. Since $\langle u_i, u_j\rangle = \overline{\langle u_j, u_i\rangle}$, the matrix of the linear system (7.2.2) is the conjugate transpose of the Gram matrix $\Gamma(\boldsymbol{u})$. More precisely, (7.2.2) can be written in the matrix form $\Gamma^*(\boldsymbol{u})\boldsymbol{c} = \langle h, \boldsymbol{u}\rangle$ where $\boldsymbol{c}$ is the unknown n-dimensional vector with components c_k. By virtue of Theorem 1.3.1, the Gram matrix $\Gamma(\boldsymbol{f})$ is nonsingular if and only if the functions $u_1, \ldots, u_n$ are linearly independent elements of H. If this is the case, then the (unique) solution of (7.2.2) is clearly given by

$$\mathbf{c} = {\Gamma^*}^{-1}(\boldsymbol{u})\langle h, \boldsymbol{u}\rangle\,. \tag{7.2.3}$$

Now we are ready to prove the following:

Theorem 7.2.2: *Let $h \in L^2(X \times Y)$ be a function such that $h \notin \mathscr{S}_{n-1}$ for a given integer $n \geq 1$. If $\boldsymbol{f} \in L^2(X)$ and $\boldsymbol{g} \in L^2(Y)$ are n-dimensional vector functions such that $s_n = \boldsymbol{f}^*\boldsymbol{g}$ satisfies $\|h - s_n\| = \rho_n(h)$, then*

$$\boldsymbol{g}(y) = \Gamma^{-1}(\boldsymbol{f}) \int\limits_X h(x,y)\boldsymbol{f}(x)\,d\mu(x) \quad \textit{for almost all } y \in Y \tag{7.2.4}$$

and

$$\boldsymbol{f}(x) = \Gamma^{-1}(\boldsymbol{g}) \int\limits_Y \bar{h}(x,y)\boldsymbol{g}(y)\,d\lambda(y) \quad \textit{for almost all } x \in X\,. \tag{7.2.5}$$

Proof: Let h and s_n be as given and set $Y' = \{y \in Y \,:\, h(-,y) \in L^2(X)\}$. The Fubini theorem states that $\lambda(Y \setminus Y') = 0$ and that

$$\|h - s_n\| = \sqrt{\int_{Y'} \phi(y)\,d\lambda(y)}$$

where

$$\phi(y) := \int_X |h(x,y) - \boldsymbol{f}^*(x)\boldsymbol{g}(y)|^2\,d\mu(x)\,.$$

Notice that

$$\sqrt{\phi(y)} = \left\| h(-,y) - \sum_{i=1}^n \bar{f}_i(-)g_i(y) \right\|_{L^2(X)}$$

for each $y \in Y'$, which we will now keep fixed until (7.2.7). Since $h \notin \mathscr{S}_{n-1}$, Theorem 7.1.1 implies that $s_n \notin \mathscr{S}_{n-1}$, hence $\bar{f}_1, \bar{f}_2, \ldots, \bar{f}_n$ are linearly independent elements of $L^2(X)$ (otherwise the number n in $\sum_{k=1}^n \bar{f}_k g_k$ could be reduced) and $\sum_{i=1}^n \bar{f}_i(-)g_i(y)$ is their linear combinations with "constants" $g_i(y)$. Thus Principle 7.2.1 with $H = L^2(X)$, $h = h(-,y)$ and $u_i = \bar{f}_i$ $(1 \leq i \leq n)$ yields a lower estimate

$$\sqrt{\phi(y)} \geq \left\| h(-,y) - \boldsymbol{f}^*(-)\mathbf{c} \right\|_{L^2(X)} \tag{7.2.6}$$

where $\boldsymbol{c}$ is an n-dimensional constant vector given by

$$\boldsymbol{c} = \Gamma^{-1}(\boldsymbol{f}) \int_X h(x,y)\boldsymbol{f}(x)\, d\mu(x) \tag{7.2.7}$$

(see (7.2.3) and take into account that $\Gamma^*(\boldsymbol{f}^*) = \Gamma(\boldsymbol{f})$). Moreover, the equality in (7.2.6) is possible only if $\boldsymbol{g}(y) = \boldsymbol{c}$. On the other hand, it follows from Cauchy's inequality that the right-hand side of (7.2.7) determines an n-dimensional vector function $\tilde{\boldsymbol{g}}$ lying in $L^2(Y')$. If we put $\tilde{\boldsymbol{g}}(y) = \boldsymbol{0}$ for each $y \in Y \setminus Y'$, then the function $\tilde{s}_n = \boldsymbol{f}^*\tilde{\boldsymbol{g}}$ lies in $\mathscr{S}_n$ and satisfies

$$\phi(y) \geq \int_X |h(x,y) - \tilde{s}_n(x,y)|^2\, d\mu(x) \qquad \text{for any } y \in Y'\,. \tag{7.2.8}$$

Integrating (7.2.8) with respect to y yields $\|h - s_n\| \geq \|h - \tilde{s}_n\|$. However, the definition (7e) of $\rho_n(h) = \|h - s_n\|$ implies that the converse inequality holds as well. Thus we have $\|h - s_n\| = \|h - \tilde{s}_n\|$, which means that the equality in (7.2.8) holds for almost all $y \in Y'$. Hence $\boldsymbol{g} = \tilde{\boldsymbol{g}}$ almost everywhere on Y, which gives (7.2.4). Analogously, minimalizing the value of

$$\Big\| \bar{h}(x,-) - \sum_{i=1}^{n} d_i \bar{g}_i(-) \Big\|_{L^2(Y)} \qquad \text{with a fixed } x \in X$$

we establish the condition (7.2.5). □

7.3. Spectral properties

After finding the integral equations (7.2.4) and (7.2.5), it becomes clear why we now have to deal with the pair of integral operators

$$T\colon L^2(X) \to L^2(Y) \quad \text{and} \quad T^*\colon L^2(Y) \to L^2(X)$$

defined by

$$Tu(y) = \int_X h(x,y)u(x)\, d\mu(x) \quad \text{and} \quad T^*v(x) = \int_Y \bar{h}(x,y)v(y)\, d\lambda(y)\,. \tag{7.3.1}$$

It is easy to check that the operators T and T^* are linear, continuous and *adjoint* to each other, because

$$\langle Tu, v\rangle_{L^2(Y)} = \langle u, T^*v\rangle_{L^2(X)} = \int_{X\times Y} hu\bar{v}\, d(\mu \times \lambda) \tag{7.3.2}$$

for any $u \in L^2(X)$ and any $v \in L^2(Y)$.

Using the notations (7.3.1), the system (7.2.4) and (7.2.5) can be rewritten as

$$\boldsymbol{g} = \Gamma^{-1}(\boldsymbol{f})T\boldsymbol{f} \quad \text{and} \quad \boldsymbol{f} = \Gamma^{-1}(\boldsymbol{g})T^*\boldsymbol{g}\,. \tag{7.3.3}$$

Hence $\boldsymbol{f}$ and $\boldsymbol{g}$ are "separable" as follows

$$\boldsymbol{f} = \Gamma^{-1}(\boldsymbol{g})\Gamma^{-1}(\boldsymbol{f})T^*T\boldsymbol{f} \quad \text{and} \quad \boldsymbol{g} = \Gamma^{-1}(\boldsymbol{f})\Gamma^{-1}(\boldsymbol{g})TT^*\boldsymbol{g}\,. \tag{7.3.4}$$

Here the operators $T^*T\colon L^2(X) \to L^2(X)$ and $TT^*: L^2(Y) \to L^2(Y)$ are clearly self-adjoint and positively semidefinite. We recall two well-known facts about such operators (cf. [Ta, pages 331–332]):
(i) each nonzero eigenvalue is real and positive;
(ii) each two eigenvectors that correspond to a pair of different eigenvalues are orthogonal.

Before stating further properties (iii) and (iv), let us prove the following:

Proposition 7.3.1: *Let $u_1, u_2, \ldots, u_n$ be an orthonormal system of functions in $L^2(X)$ such that*

$$T^*Tu_k \;=\; \beta_k^2 \cdot u_k \qquad (\beta_k > 0,\, 1 \le k \le n) \tag{7.3.5}$$

where T and T^* are operators (7.3.1)*, with a given function $h \in L^2(X \times Y)$. Then the functions $v_k = Tu_k$ $(1 \le k \le n)$ form an orthogonal system in $L^2(Y)$ such that $\|v_k\| = \beta_k$ $(1 \le k \le n)$. Moreover, the function $s_n = \sum_{k=1}^n \bar{u}_k v_k \in \mathscr{S}_n$ satisfies*

$$\|h - s_n\| = \sqrt{\|h\|^2 - \sum_{k=1}^n \beta_k^2}\,. \tag{7.3.6}$$

Proof: We have

$$\begin{aligned}\langle v_i, v_j\rangle_{L^2(Y)} &= \langle Tu_i, Tu_j\rangle_{L^2(Y)} = \langle u_i, T^*Tu_j\rangle_{L^2(X)} \\ &= \langle u_i, \beta_j^2 u_j\rangle_{L^2(X)} = \beta_j^2\langle u_i, u_j\rangle_{L^2(X)}\end{aligned} \qquad (i, j \in \{1, 2, \ldots, n\})$$

which proves the first part of Proposition 7.3.1. Since the functions $\bar{u}_k v_k$ $(1 \le k \le n)$ form an orthogonal system in $L^2(X \times Y)$, it follows that

$$\|s_n\|_{L^2(X\times Y)} = \sqrt{\sum_{k=1}^n \|u_k\|^2_{L^2(X)} \cdot \|v_k\|^2_{L^2(Y)}} = \sqrt{\sum_{k=1}^n \beta_k^2}\,.$$

Thus (7.3.6) is valid if $h - s_n$ and s_n are orthogonal elements of $L^2(X \times Y)$. An easy computation based on (7.3.2) gives

$$\langle h, s_n\rangle_{L^2(X\times Y)} = \int\limits_{X\times Y} h \cdot \left(\sum_{k=1}^{n} u_k \bar{v}_k\right) d(\mu \times \lambda) = \sum_{k=1}^{n} \langle u_k, T^* v_k\rangle_{L^2(X)}$$

$$= \sum_{k=1}^{n} \langle u_k, T^*T u_k\rangle_{L^2(X)} = \sum_{k=1}^{n} \langle u_k, \beta_k^2 u_k\rangle_{L^2(X)} = \sum_{k=1}^{n} \beta_k^2 .$$

Thus $\langle h, s_n\rangle = \langle s_n, s_n\rangle$, which yields $\langle h - s_n, s_n\rangle = 0$. □

Note that the formula (7.3.6) provides an important inequality

$$\sum_{k=1}^{n} \beta_k^2 \leq \text{const. } (= \|h\|^2)$$

which immediately ensures the following properties:
(iii) The set of all nonzero eigenvalues of T^*T is at most a countable set of positive reals whose elements α^2 $(\alpha > 0)$ can be enumerated so that

$$\alpha_1^2 > \alpha_2^2 > \alpha_3^2 > \dots \tag{7.3.7}$$

(iv) Each eigenspace

$$U_j = \{u \in L^2(X) \,|\, T^*Tu = \alpha_j^2 u\} \tag{7.3.8}$$

has a finite dimension $d_j = \dim U_j$ and the numbers d_j satisfy the inequality

$$\sqrt{\sum_j d_j \alpha_j^2} \leq \|h\|_{L^2(X\times Y)} . \tag{7.3.9}$$

Let us emphasize the following: We have still not proved that *the set of nonzero eigenvalues of T^*T is not vacuous* (provided that $h \notin \mathscr{S}_0$). We will do that in the next section by applying Theorems 7.1.1 and 7.2.2. Moreover, we will show that in fact the equality in (7.3.9) is always valid.

For the sake of final formulations we agree to write the sequence of the eigenvalues (7.3.7) with respect to their multiplicities d_j in the form

$$\beta_1^2 \geq \beta_2^2 \geq \beta_3^2 \geq \dots \tag{7.3.10}$$

which means that $\beta_k = \alpha_1$ $(1 \leq k \leq d_1)$, $\beta_k = \alpha_2$ $(d_1 + 1 \leq k \leq d_1 + d_2)$, etc.

7.4. Best approximations and Hilbert–Schmidt decompositions

Now we are in a position to prove our main results.

Theorem 7.4.1: *Let $h \in L^2(X \times Y)$ be a function not lying in $\mathscr{S}_{n-1}$ for some fixed integer $n \geq 1$. Let us retain the notation of Section 7.3. Then the following assertions are valid:*

(i) *The operator T^*T has at least n (not necessarily different) eigenvalues* (7.3.10).

(ii) *The distance $\rho_n(h)$ is given by the formula*

$$\rho_n(h) = \sqrt{\|h\|^2 - \sum_{k=1}^{n} \beta_k^2}\,. \tag{7.4.1}$$

(iii) *A function $s_n \in \mathscr{S}_n$ satisfies $\rho_n(h) = \|h - s_n\|$ if and only if it can be represented as $s_n = \sum_{k=1}^{n} \bar{u_k} T u_k$, where $u_1, u_2, \ldots, u_n$ is any orthonormal system of functions in $L^2(X)$ such that $T^*Tu_k = \beta_k^2 \cdot u_k$, $1 \leq k \leq n$.*

(iv) *If in addition, $h \notin \mathscr{S}_n$, then the best approximation s_n from* (iii) *is unique just in the case when $\beta_{n+1} < \beta_n$.*

Proof: Theorem 7.1.1 yields the existence of $s_n \in \mathscr{S}_n$ satisfying $\rho_n(h) = \|h - s_n\|$. Since $h \notin \mathscr{S}_{n-1}$, Theorem 7.2.2 implies that $s_n = \boldsymbol{f}^*\boldsymbol{g}$, where $\boldsymbol{f} \in L^2(X)$ and $\boldsymbol{g} \in L^2(Y)$ are n-dimensional vector functions that satisfy (7.3.3) and (7.3.4). Denote by L_n the subspace of $L_2(X)$ generated by $f_1, f_2, \ldots, f_n$, the components of $\boldsymbol{f}$. If C is any nonsingular $n \times n$ matrix, then the transformations $\boldsymbol{f} \mapsto C^*\boldsymbol{f}$ and $\boldsymbol{g} \mapsto C^{-1}\boldsymbol{g}$ do not vary the value of $s_n = \boldsymbol{f}^*\boldsymbol{g}$, because of $(C^*\boldsymbol{f})^*(C^{-1}\boldsymbol{g}) = \boldsymbol{f}^*CC^{-1}\boldsymbol{g} = \mathbf{f}^*\boldsymbol{g}$. This is why we may choose the basis $f_1, f_2, \ldots, f_n$ of L_n arbitrarily. Notice that the first equation in (7.3.4) means that L_n is an invariant subspace of T^*T and that the restriction $T^*T\big|L_n$ is a linear isomorphism. Moreover, this isomorphism is self-adjoint and positive definite. Hence $f_1, f_2, \ldots, f_n$ may be supposed to form an orthonormal basis of L_n that consists of the eigenvectors of T^*T: there exists a subsequence $\beta_{i_1}^2, \beta_{i_2}^2, \ldots, \beta_{i_n}^2$, $(1 \leq i_1 < i_2 < \ldots < i_n)$ of (7.3.10) such that $T^*Tf_k = \beta_{i_k}^2 f_k$ $(1 \leq k \leq n)$, which proves (i). Since the Gram matrix $\Gamma(\boldsymbol{f})$ is unit, it follows from (7.3.3) that $\boldsymbol{g} = T\boldsymbol{f}$. Now we can apply Proposition 7.3.1 with $u_k = f_k$ and $v_k = g_k$ to conclude that

$$\rho_n(h) = \|h - s_n\| = \|h - \boldsymbol{f}^*\boldsymbol{g}\| = \sqrt{\|h\|^2 - \sum_{k=1}^{n} \beta_{i_k}^2}\,.$$

The last equality implies that $\beta_{i_k} = \beta_k$, $1 \leq k \leq n$, because Proposition 7.3.1 easily yields the estimate

$$\rho_n(h) \le \sqrt{\|h\|^2 - \sum_{k=1}^{n} \beta_k^2}\,.$$

Hence the assertions (ii) and (iii) hold. It follows from the preceding considerations also that the best approximation s_n is unique if and only if all orthonormal n-tuples $u_1, u_2, \ldots, u_n \in L^2(X)$ satisfying $T^*Tu_k = \beta_k^2 u_k$ $(1 \le k \le n)$ are bases of the same n-dimensional subspace L_n in $L^2(X)$. Recalling the definition of β_k from the end of Section 7.3, we can see that this condition is equivalent to the following one: the subspace L_n is a direct sum of several of the first eigenspaces $U_1, U_2, \ldots$ from (7.3.8). This immediately yields (iv). □

Theorem 7.4.2 (Hilbert–Schmidt decomposition): *Let $h \in L^2(X \times Y)$ be a nonzero function (i.e. $h \notin \mathscr{S}_0$) and let*

$$\omega = \begin{cases} n & \text{if } h \in \mathscr{S}_n \setminus \mathscr{S}_{n-1} \\ \infty & \text{if } h \notin \mathscr{S}_n \quad \text{for any } n \in \mathscr{S}_n. \end{cases} \tag{7.4.2}$$

Then there exist two orthonormal systems $\{u_k\}_{k=1}^{\omega}$ and $\{v_k\}_{k=1}^{\omega}$ lying in $L_2(X)$ and $L_2(Y)$ respectively, and a non-increasing sequence $\{\beta_k\}_{k=1}^{\omega}$ of positive reals such that

$$h(x,y) = \sum_{k=1}^{\omega} \beta_k \bar{u_k}(x) v_k(y) \quad \text{for almost all } (x,y) \in X \times Y\,. \tag{7.4.3}$$

The sub-totals of the expansion from (7.4.3) *are the best L^2-approximations of the function h in the sense that*

$$\left\| h - \sum_{k=1}^{n} \beta_k \bar{u_k} v_k \right\| = \rho_n(h) = \sqrt{\|h\|^2 - \sum_{k=1}^{n} \beta_k^2} \qquad \text{for any } n < \omega\,. \tag{7.4.4}$$

Finally, the numbers β_k satisfy

$$\sum_{k=1}^{\omega} \beta_k^2 = \|h\|^2\,. \tag{7.4.5}$$

Proof: According to Theorem 7.4.1, the number ω in (7.4.2) is defined so that all the sequence (7.3.10) of the nonzero eigenvalues of T^*T is exactly $\{\beta_k^2\}_{k=1}^{\omega}$. Thus we can choose a system $\{u_k\}_{k=1}^{\omega} \subset L^2(X)$ as follows: $\{u_k\}_{k=1}^{d_1}$ is an orthonormal basis of U_1 (see (7.3.8)), $\{u_k\}_{k=d_1+1}^{d_1+d_2}$ is an orthonormal basis of $U_2, \ldots$. In view of property (ii) stated in Section 7.3, all the system $\{u_k\}_{k=1}^{\omega}$ is orthonormal in $L^2(X)$. If we put $v_k = \beta_k^{-1} T u_k$ for each k, we get a system $\{v_k\}_{k=1}^{\omega}$ which is orthonormal in $L^2(Y)$ (see Proposition 7.3.1). Furthermore, Theorem 7.4.1 ensures (7.4.4). Equality (7.4.5) easily follows from (7.4.4) and from the facts that $\rho_n(h) = 0$ for any $h \in \mathscr{S}_n$ and that $\rho_n(h) \to 0$ as $n \to \infty$. Finally, (7.4.4) and (7.4.5) yield decomposition (7.4.3). □

We shall finish this chapter with an improvement of Theorem 7.4.2 for the important case of symmetric functions.

Theorem 7.4.3: *Let* $(X, \mu) = (Y, \lambda)$. *Then each function* $h \in L^2(X \times X)$ *that satisfies*

$$h(y, x) = \bar{h}(x, y) \qquad \textit{for almost all } (x, y) \in X \times X \tag{7.4.6}$$

has a decomposition (7.4.3) *in which* $u_k = v_k$ *or* $u_k = -v_k$, *for each index k.*

Proof: Returning to the proof of Theorem 7.4.2, we need only to show that under condition (7.4.6), each eigenspace U_j has an orthonormal basis $\{u_k\}_{k=1}^{d_j}$ satisfying $Tu_k = \alpha_j u_k$ or $Tu_k = -\alpha_j u_k$, for each $k = 1, \ldots, d_j$. Condition (7.4.6) ensures that the operator T in (7.3.1) is self-adjoint ($T = T^*$) and that (as we now check) each U_j is an invariant subspace of T. Indeed, for any $u \in U_j$ we have

$$T^*T(Tu) = T^3 u = T(T^*Tu) = T(\alpha_j^2 u) = \alpha_j^2 Tu\,.$$

Hence $Tu \in U_j$. Thus the restriction $T|U_j$ is a self-adjoint linear isomorphism, which proves the existence of the above-described basis $\{u_k\}_{k=1}^{d_j}$ of the manifold U_j. (Notice that each eigenvalue α of $T|U_j$ satisfies $\alpha^2 = \alpha_j^2$, i.e. $\alpha = \pm\alpha_j$.) □

8 GEOMETRY OF THE D'ALEMBERT EQUATION

In this chapter[5], we return to the fundamental problem in our book, namely to the celebrated result of J. d'Alembert mentioned in the Introduction. Recall his assertion from the year 1747: each sufficiently smooth scalar function h of the form

$$h(x,y)=f(x)\cdot g(y) \tag{8a}$$

has to be a solution of the partial differential equation

$$\frac{\partial^2 \log h}{\partial x \partial y}=0 \tag{8b}$$

(see [d'A]). Let us rewrite this equation in the form which does not require the existence of the logarithm

$$\begin{vmatrix} h & h_y \\ h_x & h_{xy} \end{vmatrix} = 0 \tag{8c}$$

and call it the *d'Alembert equation*. Following [PR 1], we apply a modern geometric theory of partial differential equations to equation (8c). In this way we verify once more the fact (first observed by Th. M. Rassias in the form of a counter-example [Ra 1]) that the equation of J. d'Alembert has some solutions which do not allow any (global) decomposition (8a). For the convenience of the readers who are not specialists in this area of modern geometry, we start our considerations by recalling some basic notions. Geometric tools have obviously been of particular importance to the understanding of the structure of partial differential equations on manifolds (see [PR 3]).

[5]This chapter has been written in collaboration with A. Prástaro.

8.1. Basic background from geometry of partial differential equations

Let V and W be finite-dimensional linear spaces over the field $\mathbb{K}$ ($= \mathbb{R}$ or $\mathbb{C}$). We shall denote by $\mathrm{Hom}_{\mathbb{K}}(V;W)$ the space of all $\mathbb{K}$-linear mappings from V to W. Furthermore, we shall denote by $V \otimes_{\mathbb{K}} W$, or just simply $V \otimes W$, the tensor product between V and W over the field $\mathbb{K}$ (see [Bou]). One has the canonical isomorphism over $\mathbb{K}$-linear spaces

$$\mathrm{j}: V^* \otimes W \cong \mathrm{Hom}_{\mathbb{K}}(V;W)$$

given by

$$\mathrm{j}\Big(\sum_i \alpha^i \otimes w_i\Big)(v) := \sum_i \alpha^i(v) w_i$$

for all $\alpha^i \in V^* := \mathrm{Hom}_{\mathbb{K}}(V;\mathbb{K})$, $w_i \in W$ and $v \in V$.

Let $\mathrm{T}X$ denote the tangent bundle over the n-dimensional manifold X and let T^*X be the corresponding cotangent bundle, i.e.

$$\mathrm{T}X := \bigcup_{p\in X} \mathrm{T}_p X \quad \text{and} \quad \mathrm{T}^*X := \bigcup_{p\in X} \mathrm{T}_p^* X$$

where T_pX denotes the tangent space at $p \in X$ and $\mathrm{T}_p^*X := \mathrm{Hom}_{\mathbb{K}}(T_pX;\mathbb{K})$.

Definition 8.1.1: (i) *Let X and Y be n-dimensional and m-dimensional manifolds, respectively. Let $h\colon X \to Y$ be a differentiable mapping. Denote*

$$\begin{aligned}\mathscr{D}(X,Y) := \mathrm{T}^*X \otimes \mathrm{T}Y &:= \bigcup_{(p,q)\in X\times Y} \mathrm{T}_p^*X \otimes \mathrm{T}_qY \\ &\cong \bigcup_{(p,q)\in X\times Y} \mathrm{Hom}_{\mathbb{K}}(\mathrm{T}_pX;\mathrm{T}_qY)\,.\end{aligned}$$

Then the derivative of h is the mapping $Dh\colon X \to \mathscr{D}(X,Y)$ such that

$$Dh(p) \in \mathrm{T}_p^*X \otimes \mathrm{T}_{f(p)}Y \cong \mathrm{Hom}_{\mathbb{K}}(\mathrm{T}_pX;\mathrm{T}_{f(p)}Y)$$

at each point $p \in X$.

(ii) *The space $\mathscr{D}(X,Y)$ defined above in* (i) *is called the derivative space (of first order) for mappings $X \to Y$.*

In parts (iii)–(v)*, the mapping $\pi : Y \to X$ is a given fibre bundle structure between Y and X.*

(iii) *If h is a section of π (i.e. the composition $\pi \circ h$ is the identity on X), then the values of Dh lie in a subspace of $\mathscr{D}(X,Y)$ denoted by $\mathscr{D}(Y)$ and defined by*

$$\mathscr{D}(Y) := d^*\mathscr{D}(X,Y) = \bigcup_{p\in X}\Big\{\bigcup_{q\in\pi^{-1}(p)} \mathrm{T}_p^*X \otimes \mathrm{T}_qY\Big\}$$

where $d : X \to X \times X$ is the diagonal mapping. So the following diagram is commutative:

$$\begin{array}{ccc}
\mathscr{D}(Y) := d^*\mathscr{D}(X,Y) & \longrightarrow & \mathscr{D}(X,Y) \\
\Big\downarrow & & \Big\downarrow \\
 & & X \times Y \\
 & & \Big\downarrow \mathrm{id}_X \times \pi \\
X & \xrightarrow{\quad d \quad} & X \times X
\end{array}$$

(iv) *Let us define the k-th order derivatives $D^k h := D(D^{k-1}h)$ and the spaces $\mathscr{D}^k(X,Y)$ and $\mathscr{D}^k(Y)$ by induction with respect to the order k. More precisely, we put $\mathscr{D}^0(X,Y) := X \times Y$, $\mathscr{D}^0(Y) := Y$ and define*

$$\mathscr{D}^k(X,Y) := \mathscr{D}(\mathscr{D}^{k-1}(X,Y)) \quad \textit{and} \quad \mathscr{D}^k(Y) := \mathscr{D}(\mathscr{D}^{k-1}(Y))$$

where $\mathscr{D}^{k-1}(X,Y)$ and $\mathscr{D}^{k-1}(Y)$ are considered to be fibre bundles over X, for each $k \geq 1$.

(v) *Let $\mathrm{J}\,\mathscr{D}^k(Y)$ be the subspace of $\mathscr{D}^k(Y)$ (as well of $\mathscr{D}^k(X,Y)$) whose points are exactly all derivatives of order k for sections h of π. More precisely,*

$$\mathrm{J}\,\mathscr{D}^k(Y) := \{u \in \mathscr{D}^k(Y) \mid D^k h(\pi_k(u)) = u \quad \textit{for some } h \in C^k(Y)\}$$

where $C^k(Y)$ denotes the space of C^k-sections of the fibre bundle $\pi : Y \to X$, and $\pi_k : \mathscr{D}^k(Y) \to X$ is the canonical projection. $\mathrm{J}\,\mathscr{D}^k(Y)$ is called the jet-derivative space of order k for sections of $\pi : Y \to X$.

Remarks 8.1.2: (i) Similarly we can define $\mathrm{J}\,\mathscr{D}^k(X,Y)$. Of course, the following diffeomorphism

$$\mathrm{J}\,\mathscr{D}^k(X,Y) \cong \mathrm{J}\,\mathscr{D}^k(X \times Y)$$

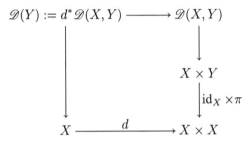

holds for any differentiable manifolds X and Y if $X \times Y$ is considered as the trivial fibre bundle π over X, i.e. $\pi(p,q) = p$ at each point $(p,q) \in X \times Y$. One has the following commutative diagram of embeddings, for each integer $k \geq 0$:

$$\begin{array}{ccc} \mathscr{D}^k(Y) & \hookrightarrow & \mathscr{D}^k(X,Y) \\ \uparrow & & \uparrow \\ \mathrm{J}\,\mathscr{D}^k(Y) & \hookrightarrow & \mathrm{J}\,\mathscr{D}^k(X,Y) \end{array}$$

(ii) As explained in [PR 1], the jet-derivative space $\mathrm{J}\,\mathscr{D}^k(Y)$ can be identified with the k-order jet-space $\mathrm{J}^k(Y)$ on the fibre bundle $\pi\colon Y \to X$, defined by

$$\mathrm{J}^k(Y) := \bigcup_{p \in X} \mathrm{J}^k(Y)_p$$

and

$$\mathrm{J}^k(Y)_p := \{ \ [h]_p^k \mid h \text{ is a section of } \pi \text{ in a neighbourhood of } p \ \}$$

where $[h]_p^k$ is the k-th equivalence class of the section h at the point $p \in X$ in the following sense: Two sections h and h' of π (defined in some neighbourhoods of p) are *equivalent in p at order k* if $D^k h(p) = D^k h'(p)$. Thus the class $[h]_p^k$ consists of all sections of π that have a contact of order k at the point p. We have the following commutative diagram:

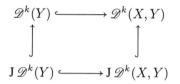

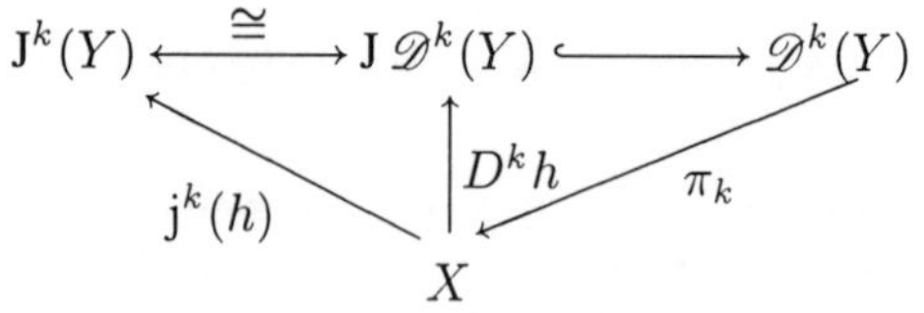

where $\mathrm{j}^k(h)$ is the k-jet of the mapping h, i.e. $\mathrm{j}^k(h)(p) := [h]_p^k$ at each $p \in X$. Of course, the embedding $\mathrm{J}\,\mathscr{D}^k(Y) \subset \mathscr{D}^k(Y)$ allows us to transfer the algebraic structure of $\mathscr{D}^k(Y)$ onto $\mathrm{J}\,\mathscr{D}^k(Y)$. Indeed, for each $u \in \mathscr{D}^{k-1}(Y)$, the fibre $\mathscr{D}^k(Y)_u$ of $\mathscr{D}^k(Y)$ over u is a linear space with the following structure

$$\mathscr{D}^k(Y)_u \cong \mathrm{T}_x^* X \otimes \mathrm{T}_u \mathscr{D}^{k-1}(Y)$$

where $x := \pi_k(u) \in X$. Thus any point $\bar{u} \in \mathrm{J}\,\mathscr{D}^k(Y)$ can be represented as a vector of the space

$$\mathrm{T}_x^* X \otimes \mathrm{T}_{\pi_{k,k-1}(\bar{u})} \mathscr{D}^{k-1}(Y).$$

In this way we can treat derivatives of mappings as *tensor fields*, i.e. we can represent them by means of differential algebraic formulas. Recall that a system of coordinates

$\{ x^i, y^j \}_{1\leq i\leq n, 1\leq j\leq m}$ on Y is called fibred on X with respect to the projection π if there exists a system of coordinates $\{ \bar{x}^i \}_{1\leq i\leq n}$ on X such that

$$x^i = \bar{x}^i \circ \pi \quad \text{for each } i = 1, 2, \ldots, n.$$

Then the coordinates $\{ y^j \}_{1\leq j\leq m}$ are called *vertical* (in contrast with x^i which are called *horizontal*). In fact if $Y_p := \pi^{-1}(p)$, $p \in X$, is the fibre of Y over p, then $\{ y^j|_{Y_p} \}_{1\leq j\leq m}$ represent coordinates on the m-dimensional manifold $Y_p \subset Y$. This consideration is very useful in order to obtain local expressions of derivatives $\mathscr{D}^k h$ of a section h of $\pi\colon Y \to X$. Then we can write the local expression of a derivative Dh of a section $h\colon X \to Y$ in the following algebraic way

$$Dh = \sum_{\substack{1\leq i\leq n,\\ 1\leq j\leq m}} \left[\delta_i^j dx^i \otimes \partial x_j + (\partial x_i \cdot h^j) dx^i \otimes \partial y_j \circ h\right]$$

where $\{ x^i, y^j \}$ are fibred coordinates on Y, y^j are the vertical coordinates, and $h^j := y^j \circ h$. Thus the derivative $Dh(p)$ is characterized by the 1-jet $\mathrm{j}(h) = \{ h^j, (\partial x_i \cdot h^j) \}$. Similarly, for $D^2 h$ we have the following algebraic representation

$$D^2 h = \sum_{\substack{1\leq i\leq n,\\ 1\leq j\leq m}} \left[\delta_i^j dx^i \otimes \partial x_j + (\partial x_i \cdot h^j) dx^i \otimes \partial y_j \circ Dh\right] + \sum_{\substack{1\leq i,s\leq n,\\ 1\leq j\leq m}} (\partial x_s \partial x_i \cdot h^j) dx^i \otimes \partial y_j^s \circ Dh\,.$$

Hence $D^2 h$ is characterized by the 2-jet $\mathrm{j}^2(h) = \{ h^j, (\partial x_i \cdot h^j), (\partial x_s \partial x_i \cdot h^j) \}$ and so on.

If any fibre Y_p, $p \in X$, of the fibre bundle $\pi : Y \to X$ has the structure of a $\mathbb{K}$-linear space, then the fibre bundle π is called a vector bundle. We shall denote a vector bundle by $\pi\colon E \to X$. Then, for each k, $\pi_k : \mathrm{J}\,\mathscr{D}^k(E) \to X$ also becomes a vector bundle:

$$D^k s_1(p) + D^k s_2(p) = D^k(s_1 + s_2)(p)$$
$$\lambda D^k s(p) = D^k(\lambda s)(p)$$

for all $\lambda \in \mathbb{K}$ and $s_1, s_2 \in C^\infty(E)$.

Definition 8.1.3: (i) *A system of partial differential equations of order k on the fibre* *bundle* $\pi\colon Y \to X$ *is a subbundle*[6] E_k *of* $\mathrm{J}\,\mathscr{D}^k(Y)$ *over* X*:*

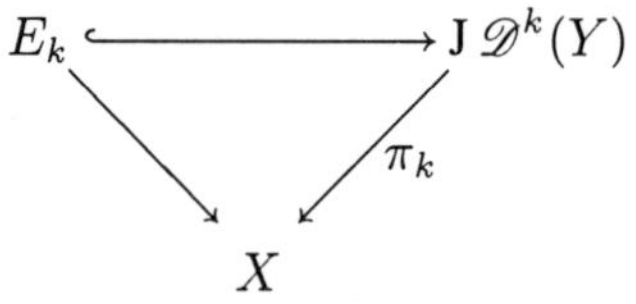

(ii) *A linear system of partial differential equations of order* k *on the vector bundle* $\pi\colon E \to X$ *is a vector subbundle* E_k *of the vector bundle* $\mathrm{J}\,\mathscr{D}^k(E) \to X$.

Remarks 8.1.4: **(i)** The local expression of E_k can be represented by means of some s-tuple of numerical functions H^i as a system

$$H^i = 0 \quad (i = 1, 2, \ldots, s) \quad \text{on } \mathrm{J}\,\mathscr{D}^k(Y)\,. \tag{E_k}$$

(ii) In the linear case, H^i in (E_k) are linear functions with respect to the vertical coordinates of $\mathrm{J}\,\mathscr{D}^k(E)$ with respect to the projection $\pi_k\colon \mathrm{J}\,\mathscr{D}^k(E) \to X$. So if

$$\{\, x^i, y^j, y^j_i, \ldots, y^j_{i_1\ldots i_k} \;\big|\; 1 \le i,\, i_1, \ldots, i_k \le n,\, 1 \le j \le m \,\}$$

are fibred coordinates on $\mathrm{J}\,\mathscr{D}^k(E)$ defined by

$$\begin{aligned} x^i(D^k h(p)) &= x^i(p) \\ y^j_{i_1\ldots i_r}(D^k h(p)) &= (\partial x_{i_1} \cdots \partial x_{i_r} \cdot h^j)(p) \quad (0 \le r \le k) \end{aligned}$$

we get

$$H^i = \sum_{\substack{1 \le j \le m,\\ 1 \le i_1, \ldots, i_r \le n}} A^{i_1\ldots i_r}_{ij}\, y^j_{i_1\ldots i_r} \quad (i = 1, 2, \ldots, s)$$

where $A^{i_1\ldots i_r}_j = A^{i_1\ldots i_r}_j(x)$ are numerical functions in the coordinates $\{\, x^i \,\}$.

Definition 8.1.5: (i) *A regular solution of* $E_k \subset \mathrm{J}\,\mathscr{D}^k(Y)$ *is a (local) section* $h\colon U \subset X \to Y$ *such that* $D^k h(U) \subset E_k$.

(ii) *The symbol of* E_k *is the family of subspaces* $g_k := \{\, g_{k,q} \,\}_{q \in E_k}$ *of the vector bundle* $p^*_{k,0}(S^0_k X \otimes \mathrm{v\,T}Y)$, *i.e.*

$$g_k := \mathrm{v\,T}E_k \cap p^*_{k,0}(S^0_k X \otimes \mathrm{v\,T}Y)$$

where $p_{k,0}$ *is the restriction of* $\pi_{k,0}\colon \mathrm{J}\,\mathscr{D}^k(Y) \to Y$ *to* E_k, *while* $S^0_k X$ *is the space of symmetric tensors of type* $(0, k)$ *on* X *and* $\mathrm{v\,T}Y$ *denotes the vertical tangent space*

[6] *Subbundle* is synonymous for *fibred submanifold*.

of $\pi : Y \to X$, *i.e.* $\mathrm{v\,T}Y := \bigcup_{q\in Y} \mathrm{T}_q Y_{\pi(q)}$, *where* $Y_{\pi(q)}$ *is the fibre of* Y *passing through* q, *i.e.* $Y_{\pi(q)} := \pi^{-1}(\pi(q)) \subset Y$. *Furthermore,*

$$S_k^0 X \otimes \mathrm{v\,T}Y := \bigcup_{q\in Y} S_k^0(\mathrm{T}_{\pi(q)} X) \otimes \mathrm{v\,T}_q Y$$

is considered to be a fibre bundle over Y.

Remark 8.1.6: If E_k is locally characterized by system (E_k), then $g_{k,q}$ is the set of vectors

$$v_q = \sum_{\substack{1\le j\le m,\\ 1\le \gamma_1,\dots,\gamma_k\le n}} X^j_{\gamma_1\dots\gamma_k}(q)\partial y_j^{\gamma_1\dots\gamma_k}(q)$$

satisfying

$$\sum_{\substack{1\le j\le m,\\ 1\le \gamma_1,\dots,\gamma_k\le n}} (\partial y_j^{\gamma_1\dots\gamma_k}\cdot H^i)(q) X^j_{\gamma_1\dots\gamma_k}(q) = 0\,.$$

Definition 8.1.7: *A PDE* E_k *is called involutive if the Spencer cohomology associated with* g_k *vanishes* (cf. [Gos, KLV], where some discussion on the Spencer cohomology is presented).

There is a workable criterion which allows us to recognize if g_k is involutive.

Proposition 8.1.8 [GS]: *Let* V *and* W *be finite-dimensional* $\mathbb{K}$*-linear spaces and let* g *be a subspace of* $\mathrm{Hom}_{\mathbb{K}}(V;W) \cong V^*\otimes W$. *Let us put*

$$g^{(1)} := (S_2^0 V\otimes W)\cap (V^*\otimes g) \subset \mathrm{Hom}_{\mathbb{K}}(V;g)$$

and, proceeding inductively, define the k*-th prolongation of* g *as*

$$g^{(k)} := (g^{(k-1)})^{(1)} = (S_{k+1}^0 V\otimes W)\cap (S_k^0 V\otimes g)\,.$$

Note that as $g \subset V^*\otimes W$, *a canonical mapping from* $g\times V$ *to* W *is induced by the mapping*

$$(V^*\otimes W)\times V \longrightarrow \mathbb{K}\otimes_{\mathbb{K}} W \cong W$$

defined by

$$\Big(\gamma = \sum_i \alpha_i\otimes w_i,\, v\Big) \mapsto \sum_i \alpha_i(v)\otimes w_i$$

for all $\alpha_i \in V^*$, $w_i\in W$ *and* $v\in V$. *Further, if* $(v_1,\dots,v_d)$ *is a basis of* V, *let us put* $g_0 := g$ *and*

$$g_i := \big\{\gamma\in g \,\big|\, \gamma(v_1) = \gamma(v_2) = \dots = \gamma(v_i) = 0\big\}$$

for each $i = 1, 2, \ldots, d-1$. *Then*

$$\dim g^{(2)} \leq \sum_{i=0}^{d-1} \dim g_i \,.$$

If the equality holds, then g *is involutive.*

Remark 8.1.9: To each k-order PDE $E_k \subset \mathrm{J}\,\mathscr{D}^k(Y)$, we can associate its prolongation of order $h \geq 0$:

$$E_{k+h} := \mathrm{J}\,\mathscr{D}^h(E_k) \cap \mathrm{J}\,\mathscr{D}^{k+h}(Y)\,.$$

Of course, in general, E_{k+h} is simply a subset, but not a submanifold of $\mathrm{J}\,\mathscr{D}^{k+h}(Y)$. Therefore, E_{k+h} does not strictly define a PDE of order $k+h$. Thus we can give the following definition.

Definition 8.1.10: *We say that a PDE* E_k *is formally integrable if, for each integer* $h \geq 0$,

(i) g_{k+h+1} *is a vector bundle over* E_k,
(ii) *the mapping* $p_{k+h+1,k+h} : E_{k+h+1} \to E_{k+h}$, *i.e. the restriction of the mapping* $\pi_{k+h+1,k+h}$ *to* E_{k+h+1}, *is surjective.*

The following theorem is useful to recognize the formal integrability by a finite number of steps.

Theorem 8.1.11 [Gos]**:** *Let* $E_k \subset \mathrm{J}\,\mathscr{D}^k(Y)$ *be a PDE of order* k. *Then there is an integer* $k_0 \geq k$ *(depending only on* k *and on the dimension of* Y*) such that*

(i) E_{k_0} *is involutive,*
(ii) *if* g_{k+h+1} *is a vector bundle over* E_k *and* $p_{k+h+1,k+h}$ *is surjective for each* $h = 0, 1, \ldots, k_0 - k$, *then* E_k *is formally integrable.*

Now, in the analytic case, the formal integrability assures the local solvability, as indicated in the following

Theorem 8.1.12 [Gos]**:** *Let* E_k *be an analytic PDE on* $\pi : Y \to X$ *which is formally integrable. Then, given* $q \in E_{k+h}$, *with* $p_{k+h}(q) = p \in X$, *there exists an analytic solution* h *of* E_k *over a neighbourhood of* p *such that*

$$D^{k+h}h(p) = q\,.$$

Remark 8.1.13: In the above theorem, the point $q \in E_{k+h}$ is called an *initial condition*. As E_k is assumed to be formally integrable, an initial condition can be chosen on any prolongation E_{k+h} of E_k, $h \geq 0$.

The following results relate the existence and uniqueness of analytic solutions to boundary conditions.

Theorem 8.1.14: (A) (Cartan–Kähler) *Let $E_k \subset J\mathscr{D}^k(Y)$ be an analytic and involutive formally integrable PDE on the fibre bundle $\pi: Y \to X$, where* $\dim X = n$ *and* dimfibre $Y = m$. *Then there exists a unique solution $s \in C^\infty(Y)$ of E_k satisfying the following boundary conditions:*

(i) $D^{k-1}s(x_0) = q$, *where q is a fixed point on $\pi_{k,k-1}(E_k) \subset J\mathscr{D}^{k-1}(Y)$ and $x_0 = \pi_{k-1}(q) \in X$;*

(ii) *For the α^i_k-parametric derivatives $(\partial x_{\alpha_1} \dots \partial x_{\alpha_k} \cdot s^j)(x)$ of class*[7] $i = 1, 2, \dots, n$, *the value of*

$$(\partial x_{\alpha_1} \dots \partial x_{\alpha_k} \cdot s^j)(x) = (h^j_{\alpha_1 \dots \alpha_k})(x)$$

is fixed for all x lying in Z, where Z is a submanifold of X passing through x_0, locally characterized by the equations

$$x^{i+1} = x_0^{i+1}, \dots, x^n = x_0^n.$$

Note that

$$\alpha^i_k = \dim g^{i-1}_k - \dim g^i_k \quad (\textit{with } g^0_k := g_k),$$

where, as in Proposition 8.1.8, the n-tuple $\{v_1, \dots, v_n\}$ is a basis of $T_{\pi_k(q)}X$, g_k is the symbol of E_k and

$$g^i_k = \{\gamma \in (g_k)_{q \in E_k} \mid \gamma(v_1) = \dots = \gamma(v_i) = 0\} = (g_k)_i.$$

Note that, because in condition (ii) $(\partial x_{\alpha_1} \dots \partial x_{\alpha_k} \cdot s^j)(x)$ *are considered to be parametric derivatives only, their values can be arbitrarily fixed by fixing $(h^j_{\alpha_1 \dots \alpha_k})(x)$. Of course the nonparametric derivatives in the equations E_k will be expressed accordingly.*

(B) (Cauchy–Kowalevski) *Let $E_1 \subset J\mathscr{D}(Y)$ be a first order involutive formally integrable analytic system on a fibre bundle $\pi: Y \to X$, where* $\dim X = n$ *and* dimfibre $Y = m$. *Assume also that $\pi_{1,0}: E_1 \to Y$ is an epimorphism and that*

$$\beta^1_1 = 0, \dots, \beta^{n-1}_1 = 0,\ \beta^n_1 = m, \textit{ where } \beta^i_1 := m - \alpha^i_1$$

with α^i_1 defined as in part (A) *for* $k = 1$. *Then there exists a unique local solution $s = (s_1, \dots, s_m)$ of E_1 such that $s|_Z = h$ on the hypersurface $Z \subset X$ locally*

[7]The components of class i are those where there do not exist indices $1 \le i_\alpha < i$, but there exist indices $i_\alpha = i$, and eventually there exist indices $i_\alpha > i$.

characterized by the equation $x^n = const.$, *for a given section* $h\colon Z \to Y|_Z$ *of* $Y|_Z$ *over* Z. *(Here* $\{x^1, \dots, x^n\}$ *is a system of coordinates on* X.*)*

The study of singular solutions is reconduced to the Cartan theory of differential systems by soldering the Cartan approach to this formal one.

Definition 8.1.15: *The Cartan distribution on* $E_k \subset \mathrm{J}\mathscr{D}^k(Y)$ *is the subbundle* $\mathbb{E}_k \subset \mathrm{T}E_k$ *spanned by tangent spaces to graphs of k-derivatives of sections of the bundle* $\pi\colon Y \to X$ *that are solutions of* E_k.

Remark 8.1.16: $\mathbb{E}_k$ is the annihilator of the module

$$\mathrm{C}\,\Omega^1(E_k) := \left\{ \alpha \in \Omega^1(E_k) \;\middle|\; (D^k s)^*\alpha = 0 \text{ for each } s \in C^\infty_{loc}(Y) \right\}$$

where $\Omega^1(E_k)$ is the module of differential 1-forms on E_k.

Definition 8.1.17: *A (singular) solution of* $E_k \subset \mathrm{J}\mathscr{D}^k(Y)$ *is meant to be an* n*-dimensional integral manifold of the corresponding Cartan distribution* $\mathbb{E}_k$, *where* n *is the dimension of* X, *the base of the fibre bundle* $\pi : Y \to X$. *We will denote by* $\mathrm{Sol}(E_k)$ *the set of all such integral manifolds and we will call it the set of solutions of* E_k.

Remarks 8.1.18: **(i)** It is useful to extend the above definition of the set of solutions of a PDE, by including in $\mathrm{Sol}(E_k)$ all the n-dimensional integral manifolds of all the prolongations E_{k+h} of E_k, $h \geq 0$, defined in Remark 8.1.9.

(ii) Regular solutions of E_k are integral manifolds of $\mathbb{E}_k$, too.

(iii) A formal theory of singular solutions of PDE's has been introduced by A. M. Vinogradov and then developed by his co-workers in Moscow, V. Lychagin and I. S. Krasil'shchick, see [KLV]. Singular solutions have also been proved important in order to interpret the meaning of quantization of PDE's and tunnelling phenomena [Pr 3].

8.2. Solutions of the d'Alembert equation

Now we are in a position to prove:

Theorem 8.2.1 [PR 1]: *The set* Sol (d'A) *of all solutions h of the d'Alembert equation*

$$\begin{vmatrix} h & h_y \\ h_x & h_{xy} \end{vmatrix} = 0\,, \qquad \text{(d'A)}$$

considered in regions of the (x, y) *plane* $\mathbb{R}^2$, *is larger than the set of all functions* h *that can be represented in the form* (8a).

Using the modern geometric theory described in Section 8.1 and following the recent work [PR 1], we prove that (d'A) is formally integrable on the trivial fibre bundle $\pi : \mathbb{R}^3 \to \mathbb{R}^2$. This will lead to the conclusion that for any point $q \in$ (d'A) (initial condition), we can construct a formal solution, i.e. a tower of points q_k lying in $(\text{d'A})_{+k}$, the k-th prolongation of (d'A), such that $\pi_{2+k,1+k}(q_k) = q_{k-1}$, where $\pi_{2+k,1+k}$ is the canonical projection

$$\pi_{2+k,1+k} : \mathrm{J}\,\mathscr{D}^{2+k}(\mathbb{R}^2, \mathbb{R}) \to \mathrm{J}\,\mathscr{D}^{1+k}(\mathbb{R}^2, \mathbb{R}) .$$

Here $\mathrm{J}\,\mathscr{D}^s(\mathbb{R}^2, \mathbb{R})$ denotes the s-order jet-derivative space for mappings $\mathbb{R}^2 \to \mathbb{R}$. In the analytic case this implies the complete integrability, i.e. the effective construction of analytic solutions in suitable neighbourhoods of any point $p \in \mathbb{R}^2$, with $p = \pi_2(q)$, where $\pi_2 : \mathrm{J}\,\mathscr{D}^2(\mathbb{R}^2, \mathbb{R}) \to \mathbb{R}^2$ is the canonical projection. Afterwards, we prove that (d'A) can be identified with the set of such points of $\mathrm{J}\,\mathscr{D}^2(\mathbb{R}^2, \mathbb{R})$ representable in the form $(D^2 h)(x, y) = D^2(f \cdot g)(x, y)$, with some factors $f = f(x)$ and $g = g(y)$. This obviously implies that functions $h(x, y) = f(x) \cdot g(y)$ are solutions of (d'A). On the other hand, any solution of (d'A) can be identified with two-dimensional integral manifolds of the Cartan distribution $\mathbb{E}_2 \subset \mathrm{T}(\text{d'A})$ of (d'A). Then we verify that such integral submanifolds cannot (in general) be represented as graphs of the 2-jet derivatives of functions h of type $h(x, y) = f(x) \cdot g(y)$. In this way, Theorem 8.2.1 will be established. A generalization to functions of more than two variables is mentioned in the final Remark 8.2.12.

For the proof of Theorem 8.2.1 we need the following lemmas.

Lemma 8.2.2 [Tr]: *Let M_i be a C^∞ manifold of finite dimension n_i, $i = 1, 2, \ldots, m$. Then each function $h\colon M_1 \times \ldots \times M_m \to \mathbb{R}$ of class C^∞ can be represented in the following form*

$$h = \sum_{i_1=1}^{\infty} \cdots \sum_{i_m=1}^{\infty} \alpha_{i_1 \ldots i_m} f^1_{i_1} \cdot \ldots \cdot f^m_{i_m}$$

where $\alpha_{i_1 \ldots i_m} \in \mathbb{R}$ are constants and $\{f^j_i\}_{i=1}^\infty$ are functions of class $C^\infty(M_j, \mathbb{R})$, linearly independent in M_j, for each $j = 1, 2, \ldots, m$.

Proof: As shown in [Tr], the following canonical isomorphism of linear functional spaces

$$C^\infty(M_1, \mathbb{R}) \otimes_{\mathrm{R}} \cdots \otimes_{\mathrm{R}} C^\infty(M_m, \mathbb{R}) \cong C^\infty(M_1 \times \ldots \times M_m, \mathbb{R})$$

holds. Thus each function $h \in C^\infty(M_1 \times \ldots \times M_m, \mathbb{R})$ can be represented in the form

$$h = \sum_{i_1=1}^{\infty} \cdots \sum_{i_m=1}^{\infty} \alpha_{i_1 \ldots i_m} f_{i_1} \otimes \cdots \otimes f_{i_m} = \sum_{i_1=1}^{\infty} \cdots \sum_{i_m=1}^{\infty} \alpha_{i_1 \ldots i_m} f_{i_1 \ldots i_m} \cdots f_{i_m}$$

where $\{f^j_i\}_{i=1}^\infty$ is a basis of $C^\infty(M_j, \mathbb{R})$ and $\alpha_{i_1 \ldots i_m} \in \mathbb{R}$ are suitable constants. □

Lemma 8.2.3: (i) *Let the k_j-tuple of functions $\{f_i^j\}_{i=1}^{k_j} \subset C^\infty(M_j, \mathbb{R})$ be linearly independent in the set M_j, for each $j = 1, 2, \ldots, m$. Put $K = k_1 k_2 \cdots k_m$. Then the K-tuple of functions*

$$\{f_{i_1}^1(x_1) \cdot f_{i_2}^2(x_2) \cdot \ldots \cdot f_{i_k}^k(x_k) \mid i_j = 1, \ldots, k_j,\ j = 1, \ldots, m\}$$

generates a linear subspace of $C^\infty(M_1 \times \ldots \times M_m, \mathbb{R})$ of dimension K. Any function $h \in C^\infty(M_1 \times \ldots \times M_m, \mathbb{R})$ belongs to this subspace if and only if

$$h = \sum_{i_1=1}^{k_1} \cdots \sum_{i_m=1}^{k_m} \alpha_{i_1 \ldots i_m} f_{i_1}^1 \cdot \ldots \cdot f_{i_m}^m \tag{8.2.1}$$

with some coefficients $\alpha_{i_1 \ldots i_m} \in \mathbb{R}$.

(ii) *In particular, if $m = 2$, $M_1 = M_2 = \mathbb{R}$ and $k_1 = k_2 = 1$, then each of such one-dimensional linear subspaces of $C^\infty(\mathbb{R}^2, \mathbb{R})$ is contained in* Sol (d'A).

Proof: Part (i) is identical with part (ii) of Lemma 3.1.2. Part (ii) follows the fact that any function $h \in C^\infty(\mathbb{R}^2, \mathbb{R})$ of the type $h(x, y) = \alpha f^1(x) f^2(y)$, with $\alpha \in \mathbb{K}$ and $f^1, f^2 \in C^\infty(\mathbb{R}^2, \mathbb{R})$, satisfies the equation (8c). □

Notation: The d'Alembert equation is not properly a PDE in the sense of Definition 8.1.3. Indeed, the d'Alembert equation defines a subset Z_2 of $\mathrm{J}\,\mathscr{D}^2(\mathbb{R}^2, \mathbb{R})$, fibred on $\mathbb{R}^2$. In fact, the Jacobian matrix of the function

$$F := u u_{xy} - u_x u_y : \mathrm{J}\,\mathscr{D}^2(\mathbb{R}^2, \mathbb{R}) \to \mathbb{R}$$

is

$$(\partial \xi_\alpha \cdot F) = [\,0 \quad 0 \quad u_{xy} \quad -u_y \quad -u_x \quad 0 \quad u \quad 0\,]$$

where

$$(\xi^\alpha) = (x, y, u, u_x, u_y, u_{xx}, u_{xy}, u_{yy})$$

are coordinate functions on $\mathrm{J}\,\mathscr{D}^2(\mathbb{R}^2, \mathbb{R})$. Therefore, F does not have locally constant rank on all Z_2. (Thus the term *partial differential relation* could be more appropriate, cf. [Gr].) On the other hand, on the open subset

$$C_2 := u^{-1}(\mathbb{R} \setminus \{0\}) \subset \mathrm{J}\,\mathscr{D}^2(\mathbb{R}^2, \mathbb{R})$$

F has locally constant rank 1. Thus $Z_2 \cap C_2$ is a subbundle of $\mathrm{J}\,\mathscr{D}^2(\mathbb{R}^2, \mathbb{R})$ over $\mathbb{R}^2$. In the following, by abuse of notation, we shall denote by (d'A) either Z_2 or $Z_2 \cap C_2$.

Remark 8.2.4: The d'Alembert equation is not a linear PDE. More precisely, (d'A) is a seven-dimensional subbundle of $\mathrm{J}\,\mathscr{D}^2(\mathbb{R}^2, \mathbb{R})$ over $\mathbb{R}^2(x, y)$ with fibres that are manifolds of dimension 5. In fact

$$\dim(\mathrm{d'A}) = \dim \mathrm{J}\,\mathscr{D}^2(\mathbb{R}^2, \mathbb{R}) - 1 = 8 - 1 = 7\,.$$

As $\dim \mathbb{R}^2 = 2$, we get $\dim(\mathrm{d'A})_p = 7 - 2 = 5$, where $(\mathrm{d'A})_p$ is the fibre of (d'A) over $p \in \mathbb{R}^2$. Thus (d'A) is a five-dimensional submanifold of the six-dimensional linear space $\mathrm{J}\,\mathscr{D}^2(\mathbb{R}^2, \mathbb{R})_p$, the fibre of $\mathrm{J}\,\mathscr{D}^2(\mathbb{R}^2, \mathbb{R})$ over $p \in \mathbb{R}^2$. However, this does not exclude that Sol (d'A) contains subsets that are linear subspaces. This has been well pointed in the following example of a four-dimensional linear space that is contained in Sol (d'A). The set of all functions $h\colon \mathbb{R}^2 \to \mathbb{R}$ defined by

$$h(x, y) = \begin{cases} \alpha_{11} x \cdot e^{-\frac{1}{y}} + \alpha_{12} x \cdot e^{-\frac{1}{y}} + \alpha_{21} \cdot e^{-\frac{1}{y}} + \alpha_{22} e^{-\frac{1}{y}} & \text{if } y > 0 \\ 0 & \text{if } y = 0 \\ \alpha_{11} x \cdot e^{\frac{1}{y}} - \alpha_{12} x \cdot e^{\frac{1}{y}} + \alpha_{21} \cdot e^{\frac{1}{y}} - \alpha_{22} e^{\frac{1}{y}} & \text{if } y < 0 \end{cases}$$

where $\alpha_{ij} \in \mathbb{R}$ are arbitrary, form a linear subspace of $C^\infty(\mathbb{R} \times \mathbb{R}, \mathbb{R})$ of dimension 4 (the case $k_1 = k_2 = 2$), belonging to the set Sol (d'A).

Lemma 8.2.5 [PR 1]: *The d'Alembert equation* $(\mathrm{d'A}) \subset \mathrm{J}\,\mathscr{D}^2(\mathbb{R}^2, \mathbb{R})$ *is a second order PDE, formally integrable on the trivial fibre bundle* $\pi\colon \mathbb{R}^3 \to \mathbb{R}^2$ *in the sense of Definition* (8.1.10).

Proof: The symbol $(g_2)_q$ of (d'A) at $q \in (\mathrm{d'A})$ is a linear subspace of

$$S_2^0(\mathrm{T}_p X) \otimes \mathrm{v}\,\mathrm{T}_u Y \cong \mathbb{R}^2 \odot \mathbb{R}^2$$

where $X = \mathbb{R}^2$, $Y = \mathbb{R}^3$, $\odot$ denotes symmetric tensor product, vT denotes the vertical tangent functor, $p = \pi_2(q)$ and $u = \pi_{2,0}(q)$. Using the coordinate system

$$(\,x,\, y,\, u,\, u_x,\, u_y,\, u_{xx},\, u_{xy},\, u_{yy}\,) \quad \text{on } \mathrm{J}\,\mathscr{D}^2(\mathbb{R}^2, \mathbb{R})$$

and setting $F := u u_{xy} - u_x u_y$ as before, we can write

$$\mathrm{v} = U^{xx} \partial u_{xx} + U^{xy} \partial u_{xy} + U^{yy} \partial u_{yy} \in (g_2)_q \quad \text{if and only if} \quad \mathrm{v} \cdot F = 0.$$

Here U^{xx}, U^{xy} and U^{yy} denote the components of the vector v in the basis $\{\partial u_{xx},\, \partial u_{xy},\, \partial u_{yy}\}$.

Thus $\mathrm{v} \in (g_2)_q$ just in the case when $U^{xy} u = 0$, i.e. $U^{xy} = 0$. Hence

$$\dim(g_2)_q = \dim \mathbb{R}^2 \odot \mathbb{R}^2 - 1 = 3 - 1 = 2.$$

Furthermore,

$$\dim(\mathrm{d'A}) = \dim \mathrm{J}\,\mathscr{D}^2(\mathbb{R}^2,\mathbb{R}) - 1 = 8 - 1 = 7. \tag{8.2.2}$$

Now, let us consider the first prolongation $(\mathrm{d'A})_{+1}$ of (d'A)

$$(\mathrm{d'A})_{+1} \subset \mathrm{J}\,\mathscr{D}^3(\mathbb{R}^2,\mathbb{R}) : \begin{cases} F = uu_{xy} - u_x u_y = 0 \\ F_x = uu_{xyx} - u_{xx}u_y = 0 \\ F_y = uu_{xyy} - u_x u_{yy} = 0. \end{cases}$$

Then we get $\dim(\mathrm{d'A})_{+1} = \dim \mathrm{J}\,\mathscr{D}^3(\mathbb{R}^2,\mathbb{R}) - 3 = 12 - 3 = 9$. Moreover,

$$\mathrm{v} = U^{xxx}\partial u_{xxx} + U^{xxy}\partial u_{xxy} + U^{xyy}\partial u_{xyy} + U^{yyy}\partial u_{yyy} \in ((g_2)_{+1})_q$$

if and only if $\mathrm{v}\cdot F$, $\mathrm{v}\cdot F_x$ and $\mathrm{v}\cdot F_y$ are zero, i.e. $uU^{xxy} = 0$ and $uU^{xyy} = 0$. Thus $\dim((g_2)_{+1})_q = 4 - 2 = 2$ and, therefore, we get

$$\dim(\mathrm{d'A})_{+1} - \dim((g_2)_{+1})_q = \dim(\mathrm{d'A}) \quad \text{as} \quad 9 - 2 = 7\,. \tag{8.2.3}$$

On the other hand, we have the following exact sequence

$$(\mathrm{d'A})_{+1} \to (\mathrm{d'A}) \underset{0}{\overset{\kappa}{\rightrightarrows}} \mathrm{F}_1 \tag{8.2.4}$$

where κ is the curvature of (d'A) (see [Gos]) and F_1 is a vector bundle on (d'A) canonically identified by means of the following commutative and exact diagrams of vector bundles over (d'A):

$$\begin{array}{ccccccccc} & & 0 & & 0 & & & & \\ & & \downarrow & & \downarrow & & & & \\ 0 & \longrightarrow & g_2 & \longrightarrow & \pi^*_{2,0}(S^0_2 X \otimes \mathrm{v}\,TY) & \longrightarrow & \mathrm{F}_0 & & \\ & & \downarrow & & \downarrow & & \| & & \\ 0 & \longrightarrow & \mathrm{v}\,\mathrm{T}(\mathrm{d'A}) & \longrightarrow & \mathrm{J}\,\mathscr{D}^2(\mathrm{v}\,TY) & \longrightarrow & \mathrm{J}\,\mathscr{D}^2(\mathrm{v}\,TY/\,\mathrm{v}\,\mathrm{T}(\mathrm{d'A})) & \longrightarrow & 0 \end{array} \tag{8.2.5}$$

and

$$0 \to (g_2)_{+1} \to \pi^*_{2,0}(S^0_3 X \otimes \mathrm{v}\,TY) \to T^*M \otimes \mathrm{F}_0 \to \mathrm{F}_1\,. \tag{8.2.6}$$

Then we have

$$\dim \mathrm{F}_1 = 2 \cdot \dim \mathrm{F}_0 - \frac{(2+2)!}{3!} + \dim((g_2)_{+1})_q$$

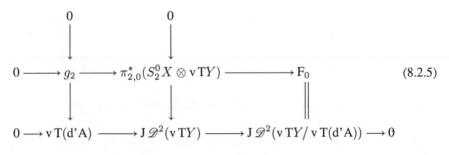

where $\dim F_0 = \dim \mathrm{J}\,\mathscr{D}^2(\mathbb{R}^2,\mathbb{R}) - \dim\,(\mathrm{d'A}) = 8 - 7 = 1$. Thus taking in account (8.2.2), we can conclude that $\dim \mathrm{F}_1 = 0$, i.e. the curvature κ is zero. Hence the canonical mapping $(\mathrm{d'A})_{+1} \to (\mathrm{d'A})$ is surjective. Now let us consider the fundamental relation (see [GS])

$$\dim((g_2)_{+1})_q \leq \sum_{i=0}^{1} \dim(g_2)_q^i$$

where $(g_2)_q^0 = (g_2)_q$ and

$$(g_2)_q^i = \left\{\, \mathrm{v} \in (g_2)_q \;\middle|\; \langle \mathrm{v}, e_1\rangle = \cdots = \langle \mathrm{v}, e_i\rangle = 0 \right\} \quad \text{for } i = 0, 1\,,$$

being $\{e_1, e_2\}$ the basis on T_pX induced by means of a coordinate system. If the equality holds, then the symbol g_2 is involutive. In fact, we have

$$\mathrm{v} = U^{xx}\partial u_{xx} + U^{yy}\partial u_{yy} \in (g_2)_q \quad \text{if and only if} \quad U^{xx} = 0\,,$$

which yields $\dim(g_2)_q^1 = 1$. Consequently,

$$\dim((g_2)_{+1})_q = 2 < 2 + 1 = \dim(g_2)_q + \dim(g_2)_q^1\,.$$

Thus (d'A) is not involutive. On the other hand, we know ([Gos]) that we can prolong an equation a finite number of times so that it becomes involutive. Indeed, this happens for $(\mathrm{d'A})_{+3}$, because the symbol $(g_2)_{+3}$ has zero dimension. Futhermore, we can see that the canonical mapping $(\mathrm{d'A})_{+4} \to (\mathrm{d'A})_{+3}$ is surjective (the proof is similar to the previous one). Therefore, we can conclude that (d'A) is formally integrable. □

Remarks 8.2.6: (i) There exist two canonical embeddings of $\mathrm{J}\,\mathscr{D}^2(\mathbb{R},\mathbb{R})$ into $\mathrm{J}\,\mathscr{D}^2(\mathbb{R}^2,\mathbb{R})$ defined by

$$a\colon \mathrm{J}\,\mathscr{D}^2(\mathbb{R},\mathbb{R}) \to \mathrm{J}\,\mathscr{D}^2(\mathbb{R}^2,\mathbb{R}), \quad (\xi,\zeta,\zeta_\xi,\zeta_{\xi\xi}) \mapsto (\xi,0,\zeta,\zeta_\xi,0,\zeta_{\xi\xi},0,0) \tag{a}$$

and

$$b\colon \mathrm{J}\,\mathscr{D}^2(\mathbb{R},\mathbb{R}) \to \mathrm{J}\,\mathscr{D}^2(\mathbb{R}^2,\mathbb{R}), \quad (\xi,\zeta,\zeta_\xi,\zeta_{\xi\xi}) \mapsto (0,\xi,\zeta,0,\zeta_\xi,0,0,\zeta_{\xi\xi})\,. \tag{b}$$

We call the embeddings a and b to be x-embedding and y-embedding, respectively. Put $A_2 := a(\mathrm{J}\,\mathscr{D}^2(\mathbb{R},\mathbb{R}))$ and $B_2 := b(\mathrm{J}\,\mathscr{D}^2(\mathbb{R},\mathbb{R}))$. Then we can identify $\mathrm{J}\,\mathscr{D}^2(\mathbb{R},\mathbb{R})$ with the two submanifolds A_2 and B_2 of $\mathrm{J}\,\mathscr{D}^2(\mathbb{R}^2,\mathbb{R})$, characterized by the following systems of equations

$$\left\{\begin{aligned} x-\xi&=0 & u_y&=0\\ y&=0 & u_{xx}-\zeta_{\xi\xi}&=0\\ u-\zeta&=0 & u_{xy}&=0\\ u_x-\zeta_\xi&=0 & u_{yy}&=0 \end{aligned}\right\} \tag{A_2}$$

and, respectively,

$$\left\{\begin{aligned} x&=0 & u_y-\zeta_\xi&=0\\ y-\xi&=0 & u_{xx}&=0\\ u-\zeta&=0 & u_{xy}&=0\\ u_x&=0 & u_{yy}-\zeta_{\xi\xi}&=0 \end{aligned}\right\} \tag{B_2}$$

(ii) The intersection $A_2 \cap B_2 \cong \mathbb{R}$ is the submanifold of $\mathrm{J}\,\mathscr{D}^2(\mathbb{R}^2,\mathbb{R})$ characterized by equations $\{x = y = u_x = u_y = u_{xx} = u_{xy} = u_{yy} = 0\}$.

Lemma 8.2.7: *The submanifolds* A_2 *and* B_2 *are properly contained into* (d'A)*:*

$$A_2 \subsetneq \text{(d'A)} \quad \textit{and} \quad B_2 \subsetneq \text{(d'A)}\ .$$

Proof: In fact, A_2 and B_2 satisfy d'Alembert equation (8c). □

Remarks 8.2.8: **(i)** There exist two canonical monomorphisms of $C^\infty(\mathbb{R},\mathbb{R})$ into $C^\infty(\mathbb{R}^2,\mathbb{R})$ defined by

$$\bar{a}(h)(x,y) = h(x) \quad \text{and} \quad \bar{b}(h)(x,y) = h(y)$$

for each $h \in C^\infty(\mathbb{R},\mathbb{R})$. We call $\bar{a}$ and $\bar{b}$ to be x-embedding and y-embedding, respectively. Set $A := \bar{a}(C^\infty(\mathbb{R},\mathbb{R}))$ and $B := \bar{b}(C^\infty(\mathbb{R},\mathbb{R}))$. Then $C^\infty(\mathbb{R},\mathbb{R})$ can be identified with two linear subspaces of $C^\infty(\mathbb{R}^2,\mathbb{R})$: A is the subspace of functions $h = h(x,y)$ that are constant with respect to the y-variable, while B is the subspace of functions $h = h(x,y)$ that are constant with respect to the x-variable. Finally, the intersection $A \cap B$, the subspace of constant functions, can be identified with $\mathbb{R}$.

(ii) A, B and $\mathbb{R}$ belong to the sets of solutions of A_2, B_2 and $A_2 \cap B_2$, respectively.

Lemma 8.2.9: (d'A) *is diffeomorphic to the image* P_2 *of the canonical mapping* $\mathrm{j} : \mathrm{J}\,\mathscr{D}^2(\mathbb{R},\mathbb{R}) \times \mathrm{J}\,\mathscr{D}^2(\mathbb{R},\mathbb{R}) \to \mathrm{J}\,\mathscr{D}^2(\mathbb{R}^2,\mathbb{R})$ *defined by*

$$\mathrm{j}(D^2 f(x), D^2 g(y)) := D^2(f \cdot g)(x,y) \quad \textit{for each } f \textit{ and } g\,.$$

More precisely, the following diagram

$$\begin{array}{ccc} \mathrm{J}\,\mathscr{D}^2(\mathbb{R},\mathbb{R})\times \mathrm{J}\,\mathscr{D}^2(\mathbb{R},\mathbb{R}) & \xrightarrow{\eta} & \mathrm{J}\,\mathscr{D}^2(\mathbb{R},\mathbb{R})\otimes \mathrm{J}\,\mathscr{D}^2(\mathbb{R},\mathbb{R}) \\ \downarrow & \searrow^{j} & \downarrow p \\ (\text{d'A}) & \hookrightarrow & \mathrm{J}\,\mathscr{D}^2(\mathbb{R}^2,\mathbb{R}) \end{array}$$

is commutative. Here η is the canonical morphism of vector bundles over $\mathbb{R}^2(x,y)$ and p is defined by the following equations

$$x\circ p = x,\ y\circ p = y,\ u\circ p = \bar{u}\bar{\bar{u}},\ u_x\circ p = \bar{u}_x\bar{\bar{u}}$$
$$u_y\circ p = \bar{u}\bar{\bar{u}}_y,\ u_{xx}\circ p = \bar{u}_{xx}\bar{\bar{u}},\ u_{yy}\circ p = \bar{u}\bar{\bar{u}}_{yy},\ u_{xy}\circ p = \bar{u}_x\bar{\bar{u}}_y$$

where $(x,\bar{u},\bar{u}_x,\bar{u}_{xx},y,\bar{\bar{u}},\bar{\bar{u}}_y,\bar{\bar{u}}_{yy})$ are coordinates on $\mathrm{J}\,\mathscr{D}^2(\mathbb{R},\mathbb{R})\times \mathrm{J}\,\mathscr{D}^2(\mathbb{R},\mathbb{R})$.

Proof: The image P_2 of j is the set of points $D^2h(x,y)\in \mathrm{J}\,\mathscr{D}^2(\mathbb{R}^2,\mathbb{R})$ that can be written in the following way:

$$D^2h(x,y) = D^2(fg)(x,y) \tag{8.2.7}$$

where $f = f(x)$ and $g = g(y)$ are C^∞-functions. Futhermore, P_2 is a seven-dimensional submanifold of $\mathrm{J}\,\mathscr{D}^2(\mathbb{R}^2,\mathbb{R})$ as p has constant rank 7. Moreover, $P_2\subseteq$ (d'A) as all points of $\mathrm{J}\,\mathscr{D}^2(\mathbb{R}^2,\mathbb{R})$ that satisfy (8.2.7) belong to (d'A), too. On the other hand, the mapping j is onto (d'A). In fact, for any point $q\in$ (d'A), we can find a point $u = ((D^2h)(x),(D^2g)(y))\in \mathrm{J}\,\mathscr{D}^2(\mathbb{R},\mathbb{R})\times \mathrm{J}\,\mathscr{D}^2(\mathbb{R},\mathbb{R})$ such that $\mathrm{j}(u) = q$. Indeed, the point u is given by $u = ((D^2h(.,y))(x),(D^2h(x,.))(y))$, where $h(x,.)$ and $h(.,y)$ are partial mappings $\mathbb{R}\to\mathbb{R}$ at the points x and y, respectively. □

Lemma 8.2.10: *The image of $C^\infty(\mathbb{R},\mathbb{R})\times C^\infty(\mathbb{R},\mathbb{R})$ into $C^\infty(\mathbb{R}^2,\mathbb{R})$ by means of the canonical mapping*

$$s: C^\infty(\mathbb{R},\mathbb{R})\times C^\infty(\mathbb{R},\mathbb{R})\to C^\infty(\mathbb{R},\mathbb{R})\otimes_{\mathbb{R}} C^\infty(\mathbb{R},\mathbb{R})\cong C^\infty(\mathbb{R}^2,\mathbb{R})$$

represents the set of solutions of (d'A) *that can be represented in the form* (8a). *Set* $\mathscr{S}_1 := s(C^\infty(\mathbb{R},\mathbb{R})\times C^\infty(\mathbb{R},\mathbb{R}))\subset$ Sol (d'A).

Proof: In fact, $h\in C^\infty(\mathbb{R}^2,\mathbb{R})$ can be expressed in the form (8a) if and only if $h\in\mathscr{S}_1$. On the other hand, it follows from Lemma 8.2.9 that if $h\in\mathscr{S}_1$, then the graph of D^2h is contained in (d'A) $\subset \mathrm{J}\,\mathscr{D}^2(\mathbb{R}^2,\mathbb{R})$. Therefore, $\mathscr{S}_1$ can be identified with the subset of Sol (d'A) defined by those integral manifolds of (d'A) that can be obtained as graphs of 2-jet-derivatives of functions $h = h(x,y)$ expressed in the form (8a). □

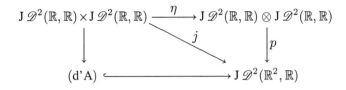

Lemma 8.2.11: *Let* $E_2 \subset J\mathscr{D}^2(M, \mathbb{R})$ *be a second-order PDE, where* M *is an* n*-dimensional manifold. Let* $F(x^\alpha, y, y_\alpha, y_{\alpha\beta}) = 0$ *be the local expression of* E_2*. Then the space of characteristic vectors at a point* $q \in E_2$ *is an* n*-dimensional space generated by the following linearly independent vector fields on* E_2*:*

$$\begin{aligned} \mathrm{v}^\beta &= (\partial y^{\alpha\beta} \cdot F)\partial x_\alpha + y_\alpha(\partial y^{\alpha\beta} \cdot F)\partial y \\ &+ y_{\delta\alpha}(\partial y^{\delta\beta} \cdot F)\partial y^\alpha - [(\partial x_\alpha \cdot F) + y_\alpha(\partial y \cdot F) + y_{\omega\alpha}(\partial y^\omega \cdot F)]\partial y^{\alpha\beta} \end{aligned}$$

Proof: The ideal $\mathscr{I}(\mathbb{E}_2)$ that characterizes the Cartan distribution $\mathbb{E}_2$ of E_2 is generated by the following differential 1-forms:

$$\begin{gathered} \omega_0 := dF = (\partial x_\alpha \cdot F)dx^\alpha + (\partial y \cdot F)dy + (\partial y^\alpha \cdot F)dy_\alpha + (\partial y^{\alpha\beta} \cdot F)dy_{\alpha\beta} \\ \omega_1 := dy - y_\alpha dx^\alpha \quad \text{and} \quad \omega_{2\alpha} := dy_\alpha - y_{\alpha\beta}dx^\beta . \end{gathered}$$

Since $X = X^\alpha\partial x_\alpha + Y\partial y + Y_\alpha\partial y^\alpha + Y_{\beta\alpha}\partial y^{\alpha\beta}$ lies in $C^\infty(\mathbb{E}_2)$ if and only if $X\rfloor_{\omega_0} = X\rfloor_{\omega_1} = X\rfloor_{\omega_2} = 0$, which implies that

$$\begin{gathered} (\partial x_\alpha \cdot F)X^\alpha + (\partial y \cdot F)Y + (\partial y^\alpha \cdot F)Y_\alpha + (\partial y^{\alpha\beta} \cdot F)Y_{\alpha\beta} = 0 \\ Y = y_\alpha X^\alpha \quad \text{and} \quad Y_\alpha = y_{\alpha\beta}X^\beta . \end{gathered}$$

X lies in Char$(\mathbb{E}_2)$, the set of characteristic vector fields of $\mathbb{E}_2$, if and only if $X\rfloor d\mathscr{I}(\mathbb{E}_2) \subset \mathscr{I}(\mathbb{E}_2)$. Taking into account that $d\mathscr{I}(\mathbb{E}_2)$ is generated by the following differential forms

$$\begin{aligned} &\omega_0, \omega_1, \omega_{2\alpha}, \omega_3 := -dy_\alpha \wedge dx^\alpha \\ &\omega_{4\alpha} := d\omega_{2\alpha} = -dy_{\alpha\beta} \wedge dx^\beta \end{aligned}$$

it follows that

$$\begin{aligned} X\rfloor_{\omega_3} &= A\omega_0 + B\omega_1 + C^\alpha\omega_{2\alpha} \\ X\rfloor_{\omega_{4\alpha}} &= \bar{A}_\alpha\omega_0 + \bar{B}_\alpha\omega_1 + \bar{C}_\alpha^\beta\omega_{2\beta}, \end{aligned}$$

which implies that

$$\begin{aligned} &dx^\alpha[-Y_\alpha - A(\partial x_\alpha \cdot F) + By_\alpha + C^\delta y_{\delta\alpha}] - dy[A(\partial y \cdot F) - B] \\ &\qquad + dy_\alpha[X^\alpha - A(\partial y^\alpha \cdot F) - C^\alpha] - dy_{\alpha\beta}A(\partial y^{\alpha\beta} \cdot F) = 0 . \end{aligned} \tag{8.2.8}$$

Thus

$$A = 0 \Rightarrow C^\alpha = X^\alpha \Rightarrow B = 0 \Rightarrow Y_\alpha = C^\delta y_{\delta\alpha} \Rightarrow Y_\alpha = X^\delta y_{\delta\alpha} .$$

Therefore, we have

$$dx^\beta[-Y_{\alpha\beta} - \bar{A}_\alpha(\partial x_\beta \cdot F) + \bar{B}_\alpha y_\beta + \bar{C}_\alpha^\omega y_{\omega\beta}] - dy[\bar{A}_\alpha(\partial y \cdot F) - \bar{B}_\alpha] \\ - dy_\beta[\bar{A}_\alpha(\partial y^\beta \cdot F) - \bar{C}_\alpha^\beta] + dy_{\omega\beta}[X^\beta \delta_\alpha^\omega - \bar{A}_\alpha(\partial y^{\omega\beta} \cdot F)] = 0\,,$$

which implies that

$$\begin{aligned} \bar{B}_\alpha &= -\bar{A}_\alpha(\partial y \cdot F) \\ \bar{C}_\alpha^\beta &= -\bar{A}_\alpha(\partial y^\beta \cdot F) \\ X^\beta \delta_\alpha^\omega &= \bar{A}_\alpha(\partial y^{\omega\beta} \cdot F)\,. \end{aligned} \tag{8.2.9}$$

This implies that $Y_{\alpha\beta} = -\bar{A}_\alpha[(\partial x_\beta \cdot F) + y_\beta(\partial y \cdot F) + y_{\omega\beta}(\partial y^\omega \cdot F)]$. Therefore,

$$\begin{aligned} X^\alpha &= \bar{A}_\beta(\partial y^{\alpha\beta} \cdot F) \\ Y &= \bar{A}_\beta y_\alpha(\partial y^{\alpha\beta} \cdot F) \\ Y_\alpha &= \bar{A}_\beta y_{\delta\alpha}(\partial y^{\delta\beta} \cdot F)\,. \quad \square \end{aligned}$$

Proof of Theorem 8.2.1: Now we are able to prove that the set of solutions of (d'A) is larger than the set $\mathscr{S}_1$ defined above in Lemma 8.2.9. In fact, the Cartan distribution $\mathbb{E}_2 \subset$ T(d'A) on (d'A) that characterizes the solutions of (d'A), is the annihilator of the ideal generated by the following differential 1-forms on J $\mathscr{D}^2(\mathbb{R}^2, \mathbb{R})$:

$$\omega_\alpha := \begin{cases} \omega_0 = dF = du\, u_{xy} + u\, du_{xy} - du_x u_y - u_x du_y \\ \omega_1 = du - u_x\, dx - u_y dy \\ \omega_2 = du_x - u_{xx}\, dx - u_{xy} dy \\ \omega_3 = du_y - u_{yx}\, dx - u_{yy} dy\,. \end{cases}$$

Let us consider two integral one-dimensional manifolds V_1 and V_2 of A_2 and B_2, respectively. Of course, we have $V_1 \subset A_2 \subset$ (d'A), $V_2 \subset B_2 \subset$ (d'A). Furthermore, both V_i are also one-dimensional integral submanifolds of (d'A). Let v_i be a characteristic vector field not contained in the tangent space TV_i, for $i \in \{1, 2\}$. More precisely, we can define

$$\mathrm{v}_1 := \mathrm{v}^x = u[\partial y + u_y \partial u + u_{yx} \partial u_x + u_{yy} \partial u_y] + u_{xx} u_y \partial u_{xx} + u_{yy} u_x \partial u_{yx} \\ \mathrm{v}_2 := \mathrm{v}^y = u[\partial x + u_x \partial u + u_{xx} \partial u_x + u_{xy} \partial u_y] + u_{xx} u_y \partial u_{xy} + u_{yy} u_x \partial u_{yy}.$$

Let $\Phi_{i,t}$: (d'A) $\to$ (d'A) be the one-parameter groups of diffeomorphisms generated by v_i: $\partial\Phi_i = \mathrm{v}_i$. Then $N_i := \bigcup_t \Phi_{i,t}(V_i)$, where $i \in \{1, 2\}$, are two-dimensional integral manifolds of (d'A). Such manifolds cannot, in general, be represented as the graphs of the second-order jet-derivatives of functions $h = h(x, y)$ of the type

(8a). In fact, let us consider, for example, the vector field v_1. The associated flow on (d'A) satisfies the following system of ordinary differential equations of first order

$$\begin{aligned}
\dot{x} &= 0 && \text{(a)}\\
\dot{y} &= u && \text{(b)}\\
\dot{u} &= u\,u_y && \text{(c)}\\
\dot{u}_x &= u\,u_{yx} && \text{(d)}\\
\dot{u}_y &= u\,u_{yy} && \text{(e)}\\
\dot{u}_{xx} &= u_{xx}u_y && \text{(f)}\\
\dot{u}_{xy} &= u_{yy}u_x && \text{(g)}\\
\dot{u}_{yy} &= 0 && \text{(h)}
\end{aligned} \tag{8.2.10}$$

together with the initial conditions at $t = 0$:

$$q_0 := (\, x_0, y_0, u_0, (u_x)_0, (u_y)_0, (u_{xx})_0, (u_{xy})_0, (u_{yy})_0 \,) \in \text{(d'A)}\ . \tag{8.2.11}$$

Let us now consider an arbitrary function $h \in C^\infty(\mathbb{R}, \mathbb{R})$ and define a manifold V by $V := D^2h(\mathbb{R}) \subset A_2 \subset$ (d'A). Thus V is an integral submanifold of (d'A) such that its tangent space does not contain the vector field v_1. Let us now fix the initial conditions in (8.2.11) for the flow associated to v_1 on such a manifold V. Then

$$N := \bigcup_t \Phi_t(V) \subset \text{(d'A)}$$

is a two-dimensional manifold of (d'A). We are interested in the projection $N_0 := \pi_{2,0}(N)$ of N on $Y := \mathbb{R}^3(x, y, u)$. In order to see what this manifold N_0 looks like, let us consider the equations in (8.2.10). From (h) we get

$$u_{yy} = (u_{yy})_0\ . \tag{i}$$

From (b) and (c) we obtain

$$\frac{du}{dy} = u_y\ . \tag{j}$$

From (b) and (e) it follows that

$$\frac{du_y}{dy} = u_{yy} = (u_{yy})_0\ . \tag{k}$$

Integrating (k) yields

$$u_y = (u_{yy})_0(y - y_0) + (u_y)_0\ . \tag{l}$$

Then from (j) and (l) we get (after integration)

$$u = \frac{(u_{yy})_O}{2}(y - y_O)^2 + (u_y)_O(y - y_O) + u_O. \tag{m}$$

Now taking in account (a), i.e. $x = x_O$, we conclude that

$$\begin{aligned} y_O &= 0 \\ u_O &= h(x) \\ (u_y)_O &= g(x) \\ (u_{yy})_O &= \ell(x) \end{aligned}$$

where h, g and ℓ are arbitrary functions in the variable x. Thus the manifold V_0 is given by the following equation

$$u(x, y) = \frac{\ell(x)}{2} \cdot y^2 + g(x) \cdot y + h(x). \tag{8.2.12}$$

By direct inspection (or by using equations (8.2.10)) we can easily conclude that $u = u(x, y)$ in (8.2.12) is a solution of (d'A) only if the functions ℓ, g and h satisfy the following equations

$$g\ell_x - \ell g_x = 0, \quad h\ell_x - \ell h_x = 0, \quad hg_x - gh_x = 0. \tag{8.2.13}$$

This is essentially equivalent to the fact that each pair of functions chosen from the triple ℓ, g and h is linearly dependent (because its Wronskian vanishes). For example, let g and ℓ be constant multiples of h: $g(x) = \alpha h(x)$ and $\ell(x) = \beta h(x)$ for some $\alpha, \beta \in \mathbb{R}$. Then (8.2.12) can be rewritten as

$$u(x, y) = \left(\frac{\beta}{2} \cdot y^2 + \alpha \cdot y + 1\right) h(x), \tag{8.2.14}$$

which of course means that u is a function of type $(8a)$. However, this conclusion follows from the fact that we have assumed that the initial manifold V is regular, i.e. the image of the second derivative of an (arbitrary) function $h \in C^\infty(\mathbb{R}, \mathbb{R})$. On the other hand, we can start from an initial singular integral manifold $V \subset A_2 \subset$ (d'A), for which there are points $q \in V$ such that

$$\operatorname{Ker} D_{\pi_2}\big|\,\mathrm{T}_q V \neq \{0\}. \tag{8.2.15}$$

Then the corresponding two-dimensional integral manifold, built up by using the above method of characteristics,

$$N := \bigcup_t V_t \tag{8.2.16}$$

has the projection N_0 on $\mathbb{R}^3(x, y, u)$ that *cannot be represented* in the form (8.2.14). *This representation is possible only in the regions outside the singular points.* For example, let us assume that V is an integral manifold of the first prolongation

$$\left\{\begin{aligned} ax^2 + bu_x^2 &= c \\ 2ax + 2bu_x u_{xx} &= 0 \end{aligned}\right\} \tag{(HJ)$_{+1}$}$$

of the following first-order equation

$$ax^2 + bu_x^2 = c \quad (a, b, c \in \mathbb{R}^+), \tag{HJ}$$

called the Hamilton–Jacobi equation. Thus V indentifies an ellipse in any plane parallel to the (x, u_x) plane and it is a cylinder in the space

$$\mathrm{J}\,\mathscr{D}(\mathbb{R}, \mathbb{R}) \cong \mathbb{R}^3(x, u, u_x)\,.$$

Of course, we have $(\mathrm{HJ})_{+1} \subset A_2 \subset (\mathrm{d'A})$. Zeros of Cartan form $\omega|(\mathrm{HJ})$ define a field of directions $\mathbb{E}(\mathrm{HJ})$ on the cylinder (HJ). The corresponding integral curves are helical ones. More precisely, $\mathbb{E}(\mathrm{HJ})$ is generated by the following vector field on V

$$\mathrm{v} := \frac{1}{u_x}\left(\partial x - \frac{a}{b}\cdot\frac{x}{u_x}\cdot\partial u_x\right) + \partial u\,.$$

Then, the corresponding integral curves satisfy

$$\left\{\begin{aligned} \dot{x} &= \frac{1}{u_x} \\ \dot{u} &= 1 \\ \dot{u}_x &= -\frac{a}{b}\cdot\frac{x}{u_x^2} \end{aligned}\right\} \tag{8.2.17}$$

Hence the parametric equations of these curves are

$$\left\{\begin{aligned} x &= \sqrt{\frac{c}{a}}\cdot\cos t \\ u &= t + \bar{u}_0 \\ u_x &= \sqrt{\frac{c}{b}}\cdot\sin t \end{aligned}\right\} \tag{8.2.18}$$

with $\bar{u}_0 \in \mathbb{R}$. The points at which v is parallel to the (x, u_x) plane (i.e. the ∂x component is zero), are characterized by the equation $1/u_x = 0$. In view of (8.2.17), this is equivalent to requiring $dx/du = 0$. Thus, from (8.2.18) we get that these points, in the (x, u) plane, are given for $u_k = \pm k\pi + u_0$, $k = 0, 1, 2, \ldots$. The projections of these curves on the (x, u) plane give a curve γ_0, where $\Sigma(\gamma_0)$ is the set of all points at which these curves are tangent to the (u_x, u) plane. *The part of the helical curves that do not project on the singular points can be represented by the derivative of a function* $h\colon \mathbb{R} \to \mathbb{R}$. As a consequence, the representation of such an integral two-dimensional manifold N and its projection N_0 cannot be done in the form (8a). In fact, N_0 can be written as $f(x, y, u) = 0$, with

$$f(x, y, u) := u - \left(\frac{\beta}{2} \cdot y^2 + \alpha \cdot y + 1\right) h_k(x) \quad \text{if } u_k \leq h_k(x) \leq u_{k+1}$$

where u_k are defined above and $u = h_k(x)$ is the part of the curve γ_0 in the (x, u) plane between $u = u_k$ and $u = u_{k+1}$. □

Remarks 8.2.12: (i) The solution (8.2.16) of the d'Alembert equation is not, in general, of type (8a). Thus in the above proof of Theorem 8.2.1, we have found a method how to construct some solutions of the d'Alembert equation different from (8a) by using integral manifolds of (d'A). It is also worth remarking that Theorem 8.2.1 interprets a general behaviour of solutions of partial differential equations. In fact, *singular solutions* or multivalued solutions cannot be interpreted as graphs of jet-derivative mappings. Therefore, a generalization of our considerations for functions in more than two variables can be, in a similar way, directly developed. In fact, it can be also proved (cf. [PR 1]) that the set of solutions of the following generalized d'Alembert equation $(\text{d'A})_k \subset J\,D^k(\mathbb{R}^k, \mathbb{R})$

$$\frac{\partial^k \log h}{\partial x_1 \partial x_2 \ldots \partial x_k} = 0$$

properly contains graphs of the k-jet derivatives of all the functions $h \in C^\infty(\mathbb{R}^k, \mathbb{R})$ that can be expressed in the form

$$h(x_1, \ldots, x_k) = h(x_1) \cdot h(x_2) \cdot \ldots \cdot h(x_k)\,.$$

(ii) In [LP], a general theory is given in order to obtain singular solutions of PDE's by using a modern geometric approach to characteristics of PDE's.

OPEN PROBLEMS

The object here is to propose some unsolved problems related to the fundamental decomposition problem that has been considered in this book.

Problem 1: J. Falmagne, a mathematical psychologist at New York University, asked in [F] about conditions on a function $h : X \times Y \to \mathbb{R}$ which guarantee the representation

$$h(x, y) = \varphi\left(\sum_{i=1}^{n} f_i(x) g_i(y)\right)$$

for all $x \in X$ and $y \in Y$ (X and Y are arbitrary sets and an unknown function $\varphi : \mathbb{R} \to \mathbb{R}$ is strictly monotonic). If φ is the identity, then Falmagne's question coincides with the fundamental decomposition problem treated in this book.

Problem 2: Let I and J be two intervals in $\mathbb{R}$ and let $n \geq 1$ be an integer. Suppose that a function $h : I \times J \to \mathbb{K}$ has the partial derivative $h_{x^n y^n}$ continuous at each point of the rectangle $I \times J$. Find a necessary and sufficient condition for the function h to be of the form $h(x, y) = f(x)g(y)$ on $I \times J$ in terms of the decomposition on the rectangle $I \times J$ into the classes

$$\mathscr{C}_r := \big\{(x, y) \in I \times J \,:\, r = \max\{j \,:\, 0 \leq j \leq n+1 \text{ and } \det W_j h(x, y) \neq 0\}\big\}$$

for $r = 0, 1, \ldots, n+1$. (Here we set $\det W_0 h(x, y) := 1$.)

Remark: This problem is motivated by the fundamental result of the book, namely by Theorem 2.1.1. In fact, this result can be now stated as follows:

(i) If h is of type $h(x, y) = f(x)g(y)$, then $\mathscr{C}_{n+1}$ is empty.
(ii) If $\mathscr{C}_{n+1}$ is empty and $\mathscr{C}_n = I \times J$, then h is of type $h(x, y) = f(x)g(y)$.

Notice that the relationship between Problem 2 and Theorem 2.1.1 resembles that between Theorem 1.2.1 and the problem solved by Wolsson in Theorem 1.2.7. In other words, to solve Problem 2, one should develop a theory of critical points for functions of two variables, analogous to Wolsson's theory for n-tuples of functions of one variable.

Problem 3: Given a function $h : X \times Y \to \mathbb{R}$ and an integer $n \geq 1$, find the best approximation

$$h(x, y) \approx \sum_{i=1}^{n} f_i(x) g_i(y)$$

with respect to the supremum norm

$$\|h\| = \sup\{|h(x, y)| \,:\, x \in X \text{ and } y \in Y\}.$$

This is even an *open problem* for polynomial functions h defined on the Cartesian product of two real intervals.

What if one changes the norm? It will be also interesting to study the best approximation problem

$$h(x_1, \ldots, x_k) \approx \sum_{i_1=1}^{m_1} \cdots \sum_{i_k=1}^{m_k} \alpha_{i_1 \ldots i_k} f_{i_1}^1(x_1) \cdot \ldots \cdot f_{i_k}^k(x_k)$$

for functions h in more than two variables, even if the L^2-norm is considered.

Remark: The corresponding problem of the best L^2-approximation (for functions of two variables only) has been solved in Chapter 7.

Problem 4: Given mnp real or complex numbers α_{ijk}, where $1 \leq i \leq m$, $1 \leq j \leq n$ and $1 \leq k \leq p$, find the smallest integer N for which the system of mnp equations

$$\sum_{s=1}^{N} a_{si} b_{sj} c_{sk} = \alpha_{ijk}$$

with unknowns a_{si}, b_{sj}, c_{sk} is solvable.

Remark: This problem is connected with the transformation of the minimal decomposition of a scalar function h of three variables

$$h(x, y, z) = \sum_{i=1}^{m} \sum_{j=1}^{n} \sum_{k=1}^{p} \alpha_{ijk} e_i(x) f_j(y) g_k(z)$$

to the diagonal form

$$h(x, y, z) = \sum_{i=1}^{N} \hat{e}_i(x) \hat{f}_i(y) \hat{g}_i(z)$$

(see Theorem 3.2.5).

Problem 5: Let $M_{m\times n}(\mathbb{K})$ denote the set of all $m \times n$ matrices with elements from the field $\mathbb{K}$ and let X and Y be two given sets. Find some necessary and sufficient conditions for a given mapping $H : X \times Y \to M_{m\times n}(\mathbb{K})$ to be factorizable into the form $H(x, y) = F(x) \cdot G(y)$, with some $F : X \to M_{m\times p}(\mathbb{K})$ and $G : Y \to M_{p\times n}(\mathbb{K})$, where p is a given positive integer. Determine also the minimal value of p for which such a factorization exists.

Problem 6: Let $X, Y_1, \dots, Y_k$ be arbitrary nonempty sets and let $GL_n(\mathbb{K})$ denote the group of all nonsingular $n \times n$ matrices with entries from the field $\mathbb{K}$. Given k surjective mappings $\varphi_i : X \to Y_i$, $1 \leq i \leq k$, find some necessary and sufficient conditions for a mapping $H : X \to GL_n(\mathbb{K})$ to be factorizable into

$$H(x) = F_1(\varphi_1(x)) \cdot F_2(\varphi_2(x)) \cdot \ldots \cdot F_k(\varphi_k(x))$$

with some factors $F_i : Y_i \to GL_n(\mathbb{K})$. We are able to solve the problem only in the case when the mapping $\varphi : X \to Y_1 \times Y_2 \times \ldots \times Y_k$ defined by

$$\varphi(x) = (\varphi_1(x), \varphi_2(x), \dots, \varphi_k(x)) \quad (x \in X)$$

is a bijection. Then the above problem can be solved by making use of the transformation $\tilde{H}(y) = H(\varphi^{-1}(y))$ to that of Section 2.4 (if $k = 2$) or Section 3.4 (if $k \geq 3$):

$$\tilde{H}(y_1, y_2, \dots, y_k) = F_1(y_1) \cdot F_2(y_2) \cdot \ldots \cdot F_k(y_k)\,.$$

Problem 7: Solve the matrix partial differential equation

$$\frac{\partial^k H}{\partial x_1 \ldots \partial x_k} = \frac{\partial H}{\partial x_1} \cdot H^{-1} \cdot \frac{\partial H}{\partial x_2} \cdot H^{-1} \cdot \ldots \cdot H^{-1} \cdot \frac{\partial H}{\partial x_k}$$

in the class of all smooth mappings $H : I_1 \times \ldots \times I_k \to GL_n(\mathbb{K})$, where $I_1, \dots, I_k$ are intervals on the real line.

Remark: Only some classes of special solutions

$$H(x_1, x_2, \dots, x_k) = H(x_1) \cdot H(x_2) \cdot \ldots \cdot H(x_k)$$

or

$$H = H(x_1, x_2, \dots, x_{i-1}, x_{i+1}, x_{i+2}, \dots, x_n)$$

are known (see Theorem 3.4.2).

REFERENCES

[AD] Aczél J. and Dhombres J., *Functional Equations in Several Variables*, Cambridge University Press, Cambridge, 1989.

[Be] Bellman R., *Introduction to Matrix Analysis*, 2nd edition, McGraw–Hill Book Company, New York, 1970.

[Bôc] Bôcher M., *Certain cases in which the vanishing of the Wronskian is a sufficient condition for linear dependence*, Trans. Amer. Math. Soc. **2** (1901), 139–149.

[Bou] Bourbaki N., *Algebra I*, Hermann, Paris, 1970.

[ČŠ 1] Čadek M. and Šimša J., *Decomposable functions of several variables*, Aequationes Math. **40** (1990), 8–25.

[ČŠ 2] Čadek M. and Šimša J., *Decomposition of smooth functions of two multidimensional variables*, Czechoslovak Math. J. **41**(116) (1991), 342–358.

[C] Curtiss D. R., *The vanishing of the Wronskian and the problem of linear dependence*, Math. Ann. **65** (1908), 282–298.

[d'A] d'Alembert J., *Recherches sur la courbe que forme une corde tendue mise en vibration I – II*, Hist. Acad. Berlin (1747), 214–249.

[F] Falmagne J., *Problem P247*, Aequationes Math. **26** (1983), 256.

[Gob] Goldberg V., *Theory of Multicodimensional $(n+1)$-Webs*, Kluwer Academic Publishers, Dordrecht, 1988.

[Gos] Goldschmidt H., *Integrability criteria for systems of nonlinear partial differential equations*, J. Diff. Geom. **1** (1967), 269–307.

[Gr] Gromov M., *Partial Differential Relations,*, Springer-Verlag, Berlin, 1986.

[GR] Gauchman H. and Rubel L. A., *Sums of products of functions of x times functions of y*, Linear Algebra Appl. **125** (1989), 19–63.

[GS] Guillemin V. and Sternberg S., *An algebraic model of transitive differential geometry*, Bull. Amer. Math. Soc. **70** (1964), 16–28.

[H] Hilbert D., *Mathematical problems* (Lecture delivered before the International Congress of Mathematicians at Paris in 1900), Bull. Amer. Math. Soc. **8** (1902), 437–479.

[He] Hille E., *Ordinary Differential Equations in the Complex Domain*, John Wiley & Sons, New York, 1976.

[Ko] Kolmogorov A. N., *On the representation of continuous functions of several variables in the form of a superposition of continuous functions of one variable and addition*, (in Russian), Dokl. Akad. Nauk. SSSR. **114** (1957), 953–956, (Amer. Math. Soc. Transl. **28** (1963), 55–59).

[Kr] Krabill D. M., *On extension of Wronskian matrices*, Bull. Amer. Math. Soc. **49** (1943), 593–601.

[KLV] Krasil'shchik I. S., Lychagin V. and Vinogradov A. M., *Geometry of Jet Spaces and Nonlinear Partial Differential Equations*, Gordon and Breach, London, 1986.

[La] Lancaster P., *Theory of Matrices*, Academic Press, New York, 1969.

[LC] Light W. A. and Cheney E. W., *Approximation Theory in Tensor Product Spaces*, Lecture Notes in Mathematics, Vol. 1169, Springer-Verlag, Berlin, 1985.

[Ly] Lychagin V., *Geometric theory of singularities of solutions of nonlinear differential equations*, J. Soviet. Math. **51**(6) (1990), 2735–2757.

[LP] Lychagin V. and Prástaro A., *Singular Cauchy data, characteristics, cocharacteristics and integral cobordism*, J. Diff. Geom. Appl. (to appear).

[MMR] Milovanović G. V., Mitrinović D. S. and Rassias Th. M., *Topics in Polynomials: Extremal Problems, Inequalities, Zeros*, World Scientific, Singapore, 1994.

[M] Montel P., *Leçons sur les recurrences et leurs applications*, Gauthier-Villars, Paris, 1957.

[N 1] Neuman F., *Functions of two variables and matrices involving factorizations*, C. R. Math. Rep. Acad. Sci. Canada **3** (1981), 7–11.

[N 2] Neuman F., *Factorizations of matrices and functions of two variables*, Czechoslovak Math. J. **32**(107) (1982), 582–588.

[N 3] Neuman F., *Functions of the form* $\sum_{i=1}^{N} f_i(x)g_i(y)$ *in* L^2, Arch. Math. (Brno) **18** (1982), 19–22.

[N 4] Neuman F., *Finite sums of products of functions in single variables*, Linear Algebra Appl. **134** (1990), 153–164.

[NR] Neuman F. and Rassias Th. M., *Functions decomposable into finite sums of products*, in *Constantin Carathéodory: An International Tribute* (ed. Rassias Th. M.), World Scientific, Singapore, 1991, pp. 956–963.

[O] Ostrowski A., *Über ein Analogon der Wronskischen Determinante bei Funktionen mehrerer Veränderlicher*, Math. Zeitschrift **4** (1919), 223–230.

[Pe] Peano G., *Sul determinante Wronskiano*, Atti Accad. Naz. Lincei Rendi. Cl. Sci. Fis. Mat. Nat. **5** (1897), 413–415.

[Pr 1] Prástaro A., *The structure of continuum systems*, Boll. Un.–Mat. Ital. **17**B (1980), 704–726, see also the same journal, Suppl. 1 (1981) pp. 69–106 and 107–129.

[Pr 2] Prástaro A., *Gauge geometrodynamics*, Riv. Nuovo Cimento **5**(4) (1982), 1–122.

[Pr 3] Prástaro A., *Quantum geometry of PDE's*, Rep. Math. Phys. **30**(3) (1991), 273–354.

[PR 1] Prástaro A. and Rassias Th. M., *A geometric approach to an equation of J. d'Alembert*, Proc. Amer. Math. Soc. (to appear).

[PR 2] Prástaro A. and Rassias Th. M., *On a geometric approach to an equation of J. D'Alembert*, in *Geometry in Partial Differential Equations* (ed. Prástaro A. and Rassias Th. M.), World Scientific, Singapore, 1994, pp. 316–322.

[PR 3] Prástaro A. and Rassias Th. M., *Geometry in Partial Differential Equations*, World Scientific, Singapore, 1994.

[PS] Pólya G. und Szegö G., *Aufgaben und Lehrsätze aus der Analysis II*, Verlag von Julius Springer, Berlin, 1925.

[Ra 1] Rassias Th. M., *A criterion for a function to be represented as a sum of products of factors*, Bull. Inst. Math. Acad. Sinica **14** (1986), 377–382.

[Ra 2] Rassias Th. M., *Problem P 286*, Aequationes Math. **38** (1989), 111–112.

[Ra 3] Rassias Th. M., *Functions decomposable into finite sums of products of one variable functions*, Abstracts, International Congress of Mathematicians, Kyoto, Japan, 1991, p. 124.

[RSz.-N] Riesz F. and Sz.-Nagy B., *Functional Analysis*, Frederick Ungar, New York, 1971.

[Ru] Rudin W., *Real and Complex Analysis*, McGraw–Hill, New York, 1966.

[Ši 1] Šimša J., *A note on certain functional determinants*, Aequationes Math. **44** (1992), 35–41.

[Ši 2] Šimša J., *The best L^2-approximation by finite sums of functions with separable variables*, Aequationes Math. **43** (1992), 248–263.

[Ši 3] Šimša J., *Approximations of two-place functions by finite sums of terms with separable variables*, J. Approx. Theory **76** (1994), 376–392.

[Ši 4] Šimša J., *Some factorizations of matrix functions in several variables*, Arch. Math. (Brno) **28** (1992), 85–94.

[Ši 5] Šimša J., *Konečné rozklady funkcí několika proměnných*, (Finite decompositions of functions of several variables, in Czech), Habilitation thesis, Masaryk University, Brno, 1991, 75 pages.

[St 1] Stéphanos C. M., *Sur une categorie d'équations fonctionelles*, Rend. Circ. Mat. Palermo **18** (1904), 360–362.

[St 2] Stéphanos C. M., *Sur une categorie d'équations fonctionelles*, in *Math. Kongress (Heidelberg, 1904)*, 1905, pp. 200–201.

[Ta] Taylor A. E., *Introduction to Functional Analysis*, John Wiley & Sons, New York, 1967.

[Tr] Treves F., *Topological Vector Spaces. Distributions and Kernels*, Academic Press, New York, 1967.

[Va] Vanžurová A., *Parallelisability conditions for differentiable three-webs*, (preprint 1994), submitted for publication.

[Vi] Vinogradov A. M., *Geometry of nonlinear differential equations*, J. Soviet. Math. **17**(1) (1981), 1624–1649.

[W 1] Wolsson K., *A condition equivalent to linear dependence for functions with vanishing Wronskian*, Linear Algebra Appl. **116** (1989), 1–8.

[W 2] Wolsson K., *Linear dependence of a function set of m variables with vanishing generalized Wronskians*, Linear Algebra Appl. **117** (1989), 73–80.

GENERAL NOTATION

$x \in X$ x is an element of X
$x \notin X$ x is not an element of X
$\{a, b, \dots\}$ set with elements $a, b, \dots$
$\{x|\dots\}$ set of all x with property ...
$\{x(t)|t \in I\}$ set of all x equal to $x(t)$ for some $t \in I$
$X \cap Y$ intersection of X and Y
$X \cup Y$ union of X and Y
$\bigcup_{i \in I} X_i$ union of all sets X_i for $i \in I$
$X \setminus Y$ set of all $x \in X$ not lying in Y
$X \subseteq Y$ X is a subset of Y
$X \subsetneq Y$ X is a proper subset of Y
$X \times Y$ Cartesian product of X and Y
$X_1 \times \dots \times X_k$ Cartesian product of sets $X_1, \dots, X_k$
X^k set of all k-tuples of elements taken from X
(x, y) pair of elements x and y
$\{f_i\}_{i=1}^n$ n-tuple $f_1, f_2, \dots, f_n$
$\mathbb{R}$ field of the real numbers
$\mathbb{C}$ field of the complex numbers
N set of all positive integers
$\mathbb{K}$ the field $\mathbb{R}$ or $\mathbb{C}$
$\mathbb{R}^+$ positive real numbers
$\mathbb{R}^k$ set of all real k-tuples
$[a, b]$ closed real interval with endpoints a and b
(a, b) open real interval with endpoints a and b
$[t_0, T)$ half-open interval of all real t satisfying $t_0 \leq t < T$
$\bar{\ }$ complex conjugation
$|x|$ the absolute value of x

$\vec{c}$ vector notation
$f : X \to Y$ mapping from X to Y
$x \mapsto f(x)$ another notation for the mapping f
$u = u(x)$ u depends on x
$u = u(x, y)$ u depends on x and y
$a \to b$ a converges to b
$f(-, y)$ or $f(\cdot, y)$ function f with respect to the first variable
$f(x, -)$ or $f(x, \cdot)$ function f with respect to the second variable
$f(x_1, \dots, x_{j-1}, -, x_{j+1}, \dots, x_n)$ function f with respect to the j-th variable
id identity function or operator
$f|_A$ restriction of the mapping f to the set A
$D_1 \circ D_2$ derivative D_1 applied to derivative D_2
f' the first derivative of f
f'' the second derivative of f
$f^{(j)}$ the j-th derivative of f
$\frac{d^j}{dx^j}$ the j-th (ordinary) derivative with respect to x
$\frac{\partial}{\partial x}$ partial derivative with respect to x
$\frac{\partial^2}{\partial x \partial y}$ second-order mixed partial derivative with respect to x and y
$\frac{\partial^k}{\partial x_1 \dots \partial x_k}$ k-th-order partial derivative with respect to $x_1, \dots, x_k$
h_x partial derivative of h with respect to x
$h_{x_i y_j}$ partial derivative of h with respect to x (i times) and to y (j times)

$A := B$ A is defined to be B
$A \approx B$ A is approximately B
$A \Rightarrow B$ B is a consequence of A
$f=g$ on X $f(x)=g(x)$ for each $x \in X$
$\log x$ logarithm of x
$\sin x$ sine of x
$\cos x$ cosine of x
$\exp x$ exponential of x
$\sum \ldots$ sum of ...
$\prod \ldots$ product of ...
$\int \ldots$ integral of ...
L^2 Hilbert space of measurable functions f with $\int |f|^2 < \infty$
$L^2(X)$ Hilbert space of functions defined on the measure space X
$\langle f, g \rangle$ inner product of f and g in L^2
$\langle \cdot, \cdot \rangle_H$ inner product in the space H
L^p Banach space of measurable functions f with $\int |f|^p < \infty$
C^n space of n-times continuously differentiable functions
$C^n(I)$ space of n-times continuously differentiable functions defined on I
$C^n(I, \mathbb{K})$ space of n-times continuously differentiable functions $f : I \to \mathbb{K}$
C^∞ space of infinitely differentiable functions
$C^\infty(X, \mathbb{K})$ space of infinitely differentiable functions $f : X \to \mathbb{K}$
(X, μ) space X with measure μ
$||h||$ norm of the function h
$||.||_H$ norm in the space H
$\max X$ maximum of X
$\sup X$ supremum of X
$\dim X$ dimension of X
$\binom{n}{k}$ binomial coefficient
$k!$ factorial of k
$[a_{ij}]_{i,j=1}^n$ $(n \times n)$ matrix with element a_{ij} in i-th row and j-th column
E_n unit matrix of order n
A^T transpose of the matrix A
A^* conjugate transpose of the matrix A
$\boldsymbol{a}^T$ row, the transpose of column $\boldsymbol{a}$
$\boldsymbol{a}^*$ row, the conjugate transpose of column $\boldsymbol{a}$
A^{-1} inverse of the matrix A
$\operatorname{rank} A$ rank of the matrix A
$\det A$ determinant of the matrix A
$U \otimes V$ tensor product of U and V
$U_1 \otimes \cdots \otimes U_k$ tensor product of spaces $U_1, \ldots, U_k$
$U \odot V$ symmetric tensor product of U and V
$A \cong B$ manifolds A and B are diffeomorphic
$X \hookrightarrow Y$ embedding X into Y
$dy_\alpha \wedge dx^\beta$ exterior product between differential forms dy_α and dx^β
$X \rfloor_\omega$ interior product between vector field X and differential form ω
$\operatorname{Ker} A$ kernel of A
$\square$ end of the proof

SYMBOL INDEX

SUBJECT INDEX